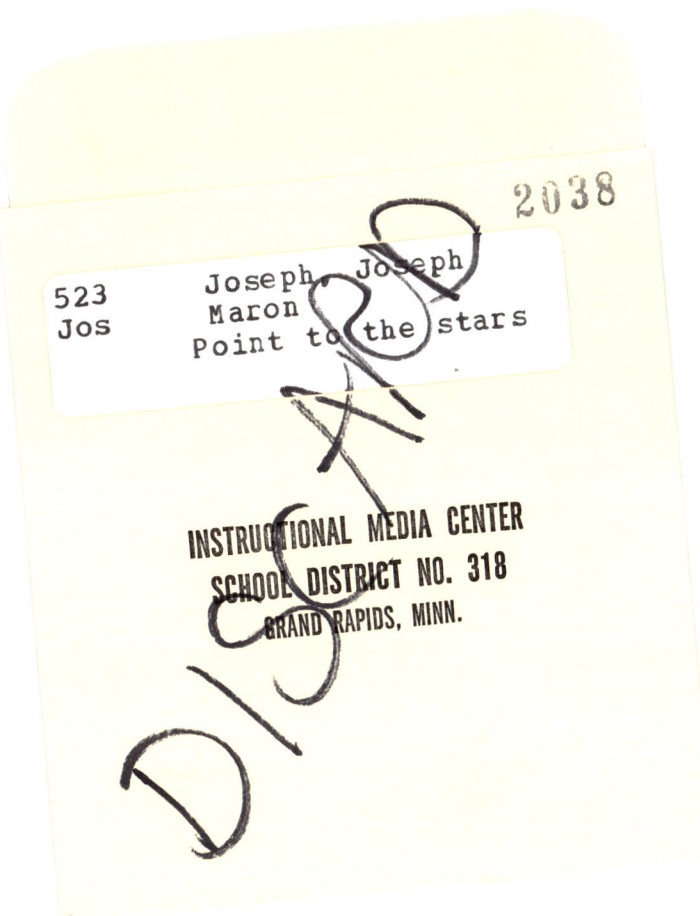

Point to the stars

revised second edition

Illustrated with photographs and line drawings

JOSEPH MARON JOSEPH
Educational Consultant

and SARAH LEE LIPPINCOTT
Director, Sproul Observatory,
Swathmore College

Point to the stars

revised second edition

INSTRUCTIONAL MEDIA CENTER
SCHOOL DISTRICT NO. 318
GRAND RAPIDS, MINN.

McGRAW-HILL BOOK COMPANY

New York • St. Louis • San Francisco • Auckland • Bogota • Düsseldorf
Johannesburg • London • Madrid • Mexico • Montreal • New Delhi • Panama
Paris • São Paulo • Singapore • Sydney • Toyko • Toronto

AUTHORS' NOTE

Calendars eventually go out of date; this is true of planet calendars also. In this revised Second Edition, the authors have provided a series of tables which will give the reader the current positions of the planets against the constellations for the next five years, through 1981. The authors have brought up to date information of artificial satellites and have included a brief description of a "black hole."

ACKNOWLEDGMENTS

Our thanks to the Mount Wilson and Palomar Observatories; the Yerkes Observatory; the late F. J. Chappell of the Lick Observatory; Jet Propulsion Laboratory, NASA; Lennart Dahlmark, science teacher, Stockholm, Sweden; and Vit-a-Way, Inc., Fort Worth, Texas, for many splendid photographs. We acknowledge the receipt of data used in the tables for planet positions from the U.S. Naval Observatory and the Fels Planetarium, Philadelphia; and of information from NASA on artificial satellites. We wish to also thank Bernard Webb, Swarthmore, Pennsylvania, for his portrayal of the "little man"; Mrs. Mary Carruth, also of Swarthmore, for her able assistance with the charts; and the U.S. Navy Hydrographic Office for the World Star Chart.

We are indebted to Julius Schwartz, specialist in science education, for his interest and enthusiasm and his many valuable suggestions during the preparation of the first edition.

Both authors extend deep appreciation to Doris Joseph for her editing and stenographic services. From the husband of this editor-at-home comes the grateful acknowledgment that woven throughout the book are a number of her excellent ideas.

Library of Congress Cataloging in Publication Data

Joseph, Joseph Maron.
   Point to the stars.

   Includes index.
   SUMMARY: Includes diagrams and text to help the reader identify stars, planets, constellations, and artificial satellites as they vary their places in the heavens.
   1. Astronomy—Observers' manuals—Juvenile literature.
[1. Constellations. 2. Stars] I. Lippincott, Sarah Lee, joint author. II. Title
QB63.J6   1977   523'.002'02   76-53737

ISBN 0-07-033050-6

Copyright © 1963, 1967, 1972, 1977 by Joseph Maron Joseph and Sarah Lee Lippincott. All Rights Reserved. Printed in the United States of America. No part of this publication may be reproduced, stored in a retrieval system, or transmitted, in any form or by any means, electronic, mechanical, photocopying, recording, or otherwise, without the prior written permission of the publisher.

3456 RABP 789

CONTENTS

Chapter I   Star light and sky sights   7

Chapter II   Pointing to the stars   18

Chapter III   Pointing to the planets   80

Chapter IV   Pointing to the artificial satellites   88

Glossary   92

Index   94

CHAPTER I

Star light and sky sights

Twinkling star lights have always caused man to look up and marvel at the beautiful sight of the night sky. Even with our vast scientific knowledge of these little points of light, we share a feeling of awe and wonderment with the ancient people who gave the bright stars the fanciful names we still use today, such as Deneb, Rigel, Algol and Arcturus.

How many stars can you see?

The stars can best be seen away from city lights—at sea or in the country. Even the casual observer occasionally looks up, notes that there are relatively few very bright stars, and that there are many dimmer stars. The stars, both bright and faint, have been counted many times by astronomers through the ages. The experienced person with normal eyesight, at a place where the sky is dark, can count about 2,500 stars during the course of a year. If your eyes are especially keen, you can count about 3,000 stars. If you wear glasses and remove them, you will be able to count less then 2,500. Scientists who specialize in the study of human eyes tell us that young people have better eyesight for seeing faint objects than their elders. If you use binoculars, or a telescope, you can see many more stars. With the 200-inch Hale telescope on Mount Palomar, in California, the largest telescope in the world, astronomers can detect more than a billion stars.

The constellations

Besides the difference in brightness you see among stars, you can distinguish different colors—certain stars always appear red and others blue-white. Observers with some imagination comment that these jewel-colored stars look as beautiful as diamonds, rubies and sapphires, which sparkle when light is played on them. It is an easy step from admiring the scintillating beauty of the night to imagining familiar objects in the patterns the stars make—such as squares, crosses and dippers. In a somewhat similar way we like to see the outline of familiar figures made by changing cloud formations.

The ancient Babylonians, Egyptians, Greeks, Romans and Hebrews saw the same stars that we see. They noticed, as we do, that the stars give the impression that they are fixed in the ceiling of a dome, or hemisphere. The observer has the feeling that he is at the center of this dome. The ancients also formed imaginative patterns with the bright stars in the sky. To them, the different designs represented their heroes, their gods and goddesses and their folk stories. The immortal ones were in their abode in this dome above the earth. From these heavenly observation posts they watched over the daily lives of mortals. This heavenly abode has been known by many names, but we use a form of the Latin word for heaven—celestial—to describe the sky dome. You can understand why the sky is called the "heavenly" or "celestial" sphere, and the beauties of the night sky "celestial objects."

Placed on the dome are the bright stars that form groups, called *constellations*. *Constellation* is the combination of two Latin words: *con*, meaning "together," and *stella*, meaning "star." Thus, "constellation" means a group of stars. The first reliable record of them comes to us from an Egyptian, Ptolemy, who knew these groups and catalogued the stars more than 1800 years ago. He indicated that they were old then.

You will notice that there are stars distributed all over the night sky, although some portions have a greater number of them than other parts. The constellations are haphazardly placed about—some large and some small. Ptolemy listed 48 constellations, and since then others have been added over the entire sky until now there are 88, which are generally known by their Latin names. For example, the group of stars called the Big Bear in English is universally known as Ursa Major.

This book contains maps which will help you to identify the various constellations visible to people in the northern hemisphere. Do not be disappointed if you do not see on the sky dome the figure corresponding to the name of the constellation. Relatively few look like the animal or figure for which they are named. It is difficult for us today to imagine why some of the star groups received the names they did.

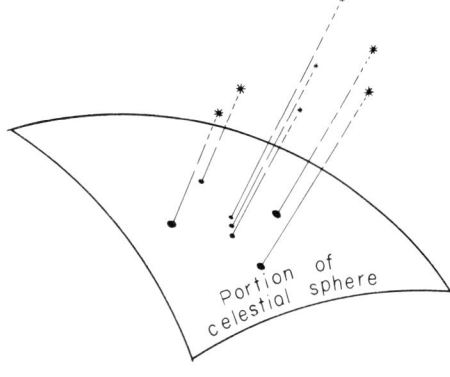

Fig. I-1. The stars appear on the celestial sphere. Stars are at different distances from the earth. They all appear projected on the celestial sphere and we cannot tell by simple inspection which ones are nearer and which are farther away from us.

But in olden times these groups served as symbols and reminded people of some important event or of some important person. Leo the Lion in the sky does not accurately resemble a lion, but when it was named, the sun was in that part of the sky in the summer, and the heat was "fierce as a lion."

From our earth view, a constellation appears to have its stars fixed on a real sky dome so that all stars are equally distant from us. In the last hundred years, astronomers have found a way to determine the distances of stars and find that some are relatively close and some much farther away. (Figure I-1.) If we were able to move to a very, very distant space station and thus get a space view, then we could see that these stars are at different depths. In fact, many of the stars in a constellation are strung out in space and have no real relation to each other.

In the daytime we cannot see the stars, but they are there. The sun lights up the earth's atmosphere, giving it a "sky blue" appearance brighter than that of any of the stars. The higher we go above the earth, the darker the daytime sky becomes. Astronauts say that when they rise beyond the greater part of the earth's atmosphere, the stars shine against a dark sky and the sun appears as a blinding ball of light.

The sun is really a star, similar to many of the points of light you see in the night sky, but it is so very much nearer that we see it as a bright, glowing sphere in the sky. Our eyes need some type of protective cover to examine the surface of the sun. The human reaction to the bright light of the sun high in the sky is to close the eyes tightly, or drop the head quickly. Even at sea level with many miles of the atmosphere intervening, we cannot look directly at the sun without

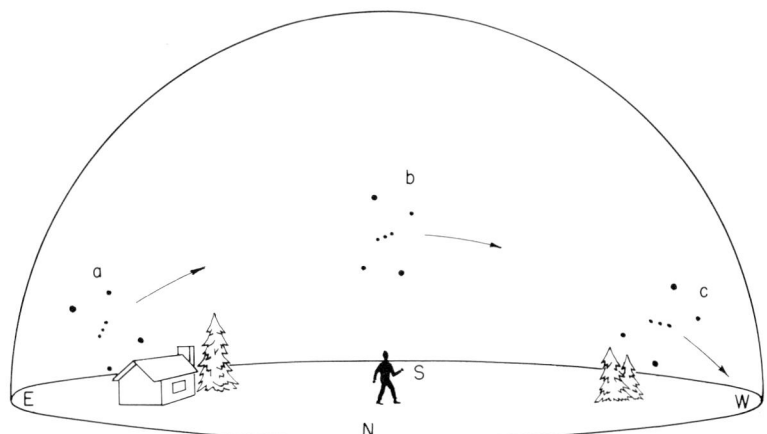

Fig. I-2. Apparent rotation of the star dome. An earth view. (**A**) Looking south. The star dome appears to rotate from east to west. Notice how Orion appears early in January: (a) early evening, (b) midnight, (c) before dawn.

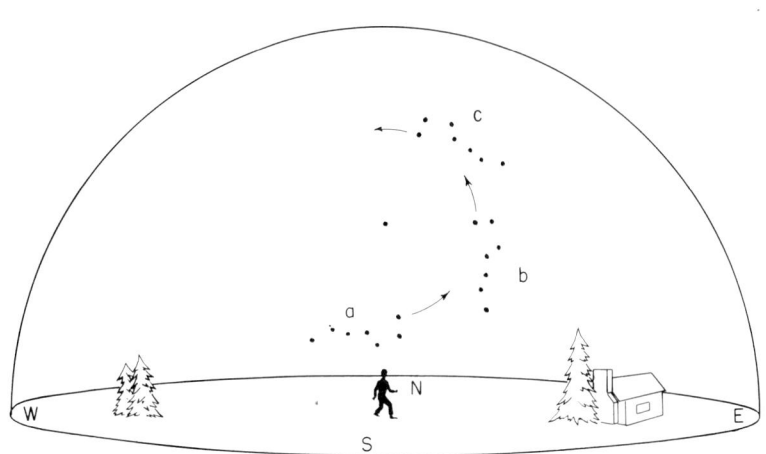

(**B**) Looking north at the same times as above.

specially prepared glasses or dense filters. (A word of caution: ordinary optical dark glasses, "sun glasses," will *not* prevent bright sunlight from injuring the eyes when you look up at the sun.)

Apparent rotation of sky dome

The stars appear fixed, that is, they hold their relative positions. The constellations do not move with respect to one another, but the whole sky dome seems to rotate during an evening. (Figure I–2.) The stars and also the sun, projected on the sky dome, appear to rise in the east and set in the west. The dome appears to rotate about an imaginary axis which extends through the earth's center, from north to south

10

Fig. I-3. Apparent rotation of celestial sphere with respect to an observer on earth.

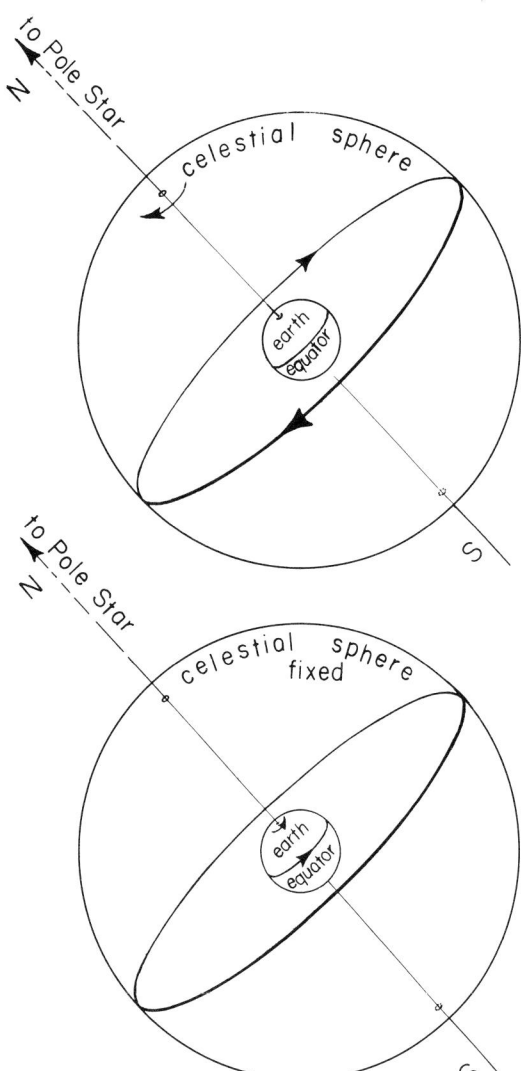

Fig. I-4. Actual rotation of earth. Earth is rotating opposite from the direction of celestial sphere. Compare with Fig. I-3.

poles, and pierces the dome close to the Pole Star, or Polaris (Figure I-3). The Pole Star appears in the same position each night and therefore does not rise or set. Stars near the Pole Star are called *circumpolar* and circle around the pole but never actually set. (See photograph of circumpolar star trails on page 45.)

The rotation of the earth

We know that the sky dome does not really rotate. A space view (Figure I-4) shows the earth rotating on its axis. Part of the time during one rotation we are looking in the direction of the sun and cannot see the stars against the daytime sky. The rotation gives us a

11

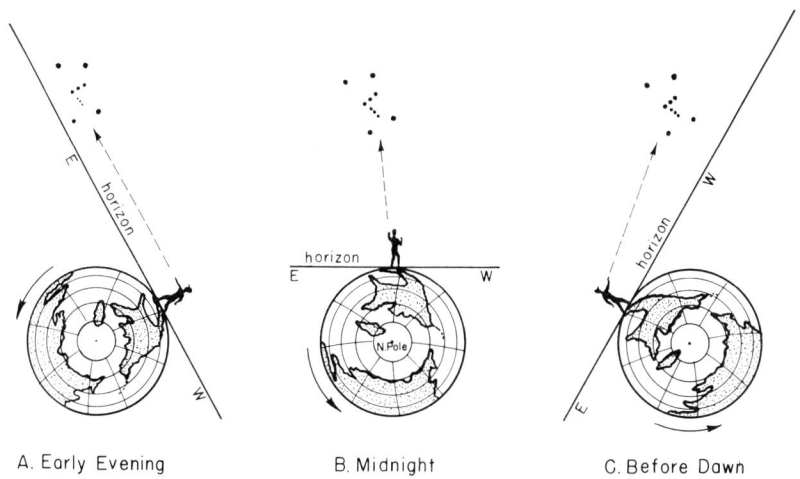

A. Early Evening   B. Midnight   C. Before Dawn

Fig. I-5. True rotation of earth, space view. The little man is standing on the equator. He always faces Orion. Orion remains fixed in space. The earth rotates so that Orion appears to rise in the east and set in the west. A, B, C correspond to a, b, c in Figure 1-2A, which is the earth view.

panoramic view of outer space, showing constellations rising in the east and setting in the west (Figure I–5).

The earth rotates on its north-south axis with a period of about 24 hours—23 hours and 56 minutes, to be exact—according to the time we keep on our watches. This is 4 minutes short of "a day." You can check this yourself in the following manner. If you observe a star (Figure I–6) passing over a landmark such as a chimney, and note the time, you will find that 23 hours and 56 minutes later the same star will pass over the same landmark, if you stand precisely on the same spot. A week later, therefore, the star will rise, or pass, over this particular landmark 7 times 4, or 28 minutes earlier.

The keen observer of the skies can detect the effect of the earth's rotation by watching the night sky for only an hour or so; stars in the west gradually disappear below the horizon while others will come into

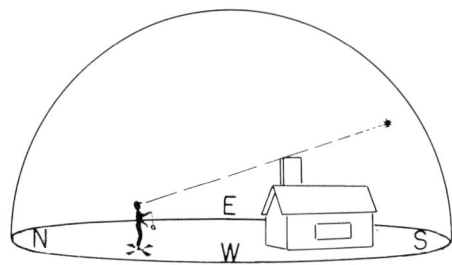

Fig. I-6. Timing the rotation period of the celestial sphere. The little man stands at a spot marked X for several successive nights. Each night the same star returns to the position just over the chimney 4 minutes earlier by his watch. In other words, the star dome makes one complete rotation in $24^h0^m - 4^m = 23^h56^m$.

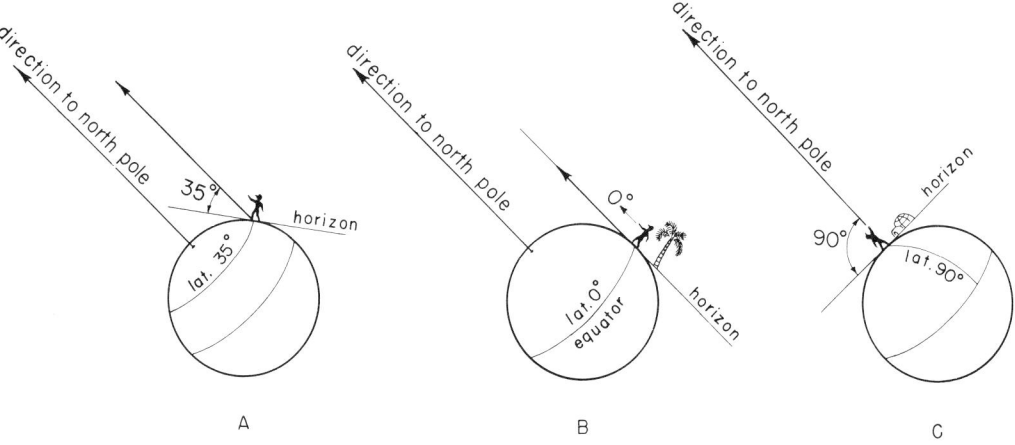

Fig. I-7. Altitude of Pole Star is equal to the latitude of the little man.
(A) For the little man at latitude +35°.
(B) For the little man at latitude 0° (Equator).
(C) For the little man at latitude +90° (North Pole).

view in the east. It is interesting to watch this effect of the earth's rotation on its axis.

The direction of the earth's axis remains fixed for many generations so that Polaris will always indicate the direction north for us. The altitude of Polaris depends on the latitude of the observer; in fact, its altitude in degrees has the same value as the observer's latitude, as shown in Figure I-7. At the equator, 0° latitude, the Pole Star appears on the horizon. Therefore, if the observer were located south of the equator he could not see Polaris; instead he could see the extension of the south pole on the celestial sphere (Figure I-4).

Changes in the night sky—the earth's revolution around the sun

Why does the sky change from month to month? The earth is in orbit; it revolves once around the sun in one year, so that during the course of a complete revolution we have a chance to view all directions of outer space unobstructed by the sun. We can think of ourselves as being on a merry-go-round. The view changes as we look out. After one revolution we have seen all around, but we cannot see in every direction at once, as the center of the carousel blocks our view just as the sun does in our earth-go-round. Since the bright sun obscures the stars in the daytime sky, we can see only the stars in that portion of the sky which is on the opposite side of the earth from the sun. From a

13

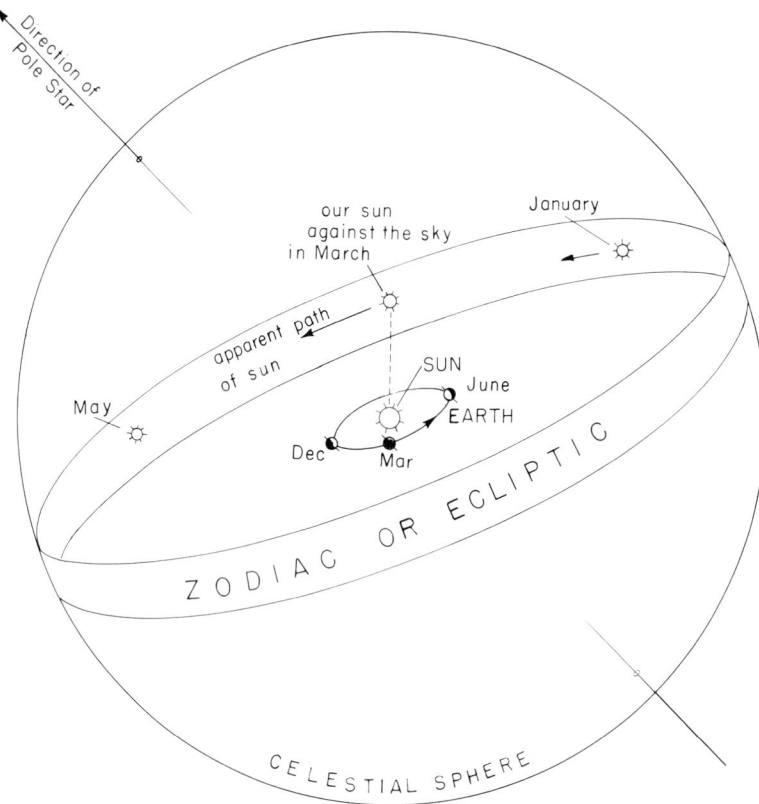

Fig. I-8. Earth-go-round. The orbit of the earth around the sun is shown. You note that the sun is fixed at the center. As the earth moves to the right from December to June, the sun appears to move to the left along the zodiac.

point out in space where not even a trace of our atmosphere can be detected, an observer will be able to see stars in all directions at the same time. There the sky is dark even near the direction of the sun, which will appear as a small blinding disk.

The summer night constellations are overhead during the winter daytime, and the autumn constellations are in the sky during the spring daytime. The opposite is also true. Later on, when you have learned more about the constellations, you can tell which ones are overhead, but invisible, in the daytime.

For a long time in history man thought that the sun revolved around the earth, but for the past several hundred years, it has been well established that the earth revolves around the sun in a well-defined orbit. The ancients were misled because the sun appears to revolve around the celestial sphere in a year (Figure I-8). From this diagram you can see why they were confused. It occurred to the ancients that a calendar could be made if they divided this path into twelve parts with a constellation assigned to each. They gave animal names to the constellations and thus the path became known as the *zodiac,* or "the path of little animals." Each animal was noted by its symbol or sign, and, when the sun had passed through all twelve signs, the people marked the passing of a year. You may have noticed these twelve symbols in farmers' almanacs. This band in the sky is known as the *ecliptic,* because the eclipses of the sun and moon occur along this path.

The moon

There are several celestial objects we have not mentioned at all which you are bound to be aware of if you have looked up at the sky for a number of nights. You probably cannot remember when you first discovered the moon, but you have to watch it on a number of consecutive nights to realize that it grows from a slim crescent, setting shortly after the sun sets, to a full moon two weeks later, rising as the sun is setting. The moon is our only natural satellite. It makes a round trip tour of the sky in a month. Can you see where our word "month" came from? The path of the moon in the sky is not far off from the one the sun follows among the constellations of the zodiac.

The planets

Among the objects visible to the naked eye in the night sky are five bright ones which do not keep with the same constellation stars. The Babylonians and Greeks also noticed these five and observed that they moved among the constellations. They called them *planets,* which means "moving stars," in order to distinguish them from the fixed stars in the constellation figures. The five planets which the ancient astronomers saw were called Mercury, Venus, Mars, Jupiter and Saturn, after their gods and goddesses. Along the Mediterranean, there are temple ruins which were used as places of worship to these planet deities. They also considered the sun and the moon to be planets, and it was not until the sixteenth century that these two were properly understood to be different from the planets.

At the present time we know that there are nine planets (including the earth), although we can see with the naked eye only the five known to the ancients. The others, Uranus, Neptune and Pluto, can be observed through a telescope. The planets are dark objects like the earth. We see them because the light from our sun is reflected from them back to us. Thus, we see our moon and the planets in much the same way that we see a stone wall illuminated by the sun. The planets are smaller than all but the smallest stars. They appear brighter because they are thousands of times nearer to us than the nearest star.

"Shooting stars" and comets

Frequently, on clear dark nights, you may see what seems to be a star that darts rapidly a short distance across the sky and then vanishes. These are popularly miscalled "shooting stars," and may look like a Fourth of July firecracker burning in a rocket stream. They are small masses of iron or stone, generally about the size of sand grains, flying through space. The majority of them burn up when they hit the earth's atmosphere, which causes them to glow for a few seconds. Astronomers call them *meteors*.

Occasionally other sights are seen in the night skies. These may be *comets*, which look like stars with tails of faint light. A few comets are bright enough to be seen with the naked eye when they pass near the earth in their orbits around the sun.

Other astronomical objects

With a telescope we can see other interesting objects. For example, there are clusters of stars which move like swarms of bees at great distances from the earth. There are also great masses of luminous gas, in cloudlike formations called *nebulae*. Scattered all over the sky are stars in pairs, called *double stars*. Occasionally, certain stars change their brightness, some in a regular way, others unexpectedly. These are *variable stars*. A dim star that suddenly bursts into great brilliance and soon fades again is known as a *nova*, or new star. All these fascinating astronomical objects will be pointed out to you in the sky when you come to Chapter II.

The Milky Way—our galaxy

The summer and autumn nights are the best times to see the Milky Way, that hazy band of light stretching across the sky. It consists of millions of distant stars and star clouds, forming a band which completely encircles the celestial sphere. Although we can see stars in all directions in the sky, there is a noticeable concentration of stars within this band. All the stars you see with your naked eye belong to our Milky Way system, or galaxy, which has a shape similar to a watch, or a very flattened sphere, or a disk. The sun and its planets are within the disk but located far from its center, so that the Milky Way band

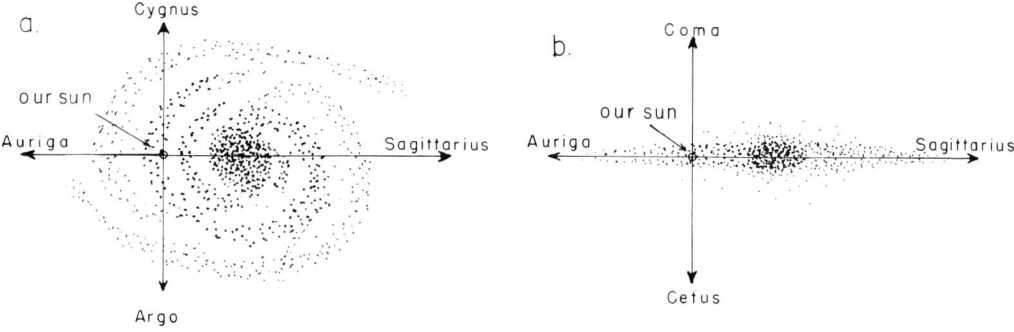

Fig. I-9. **Space view of our galaxy.** (a) Viewed from above. (b) Viewed from the side.

does not appear equally bright in all directions. It is brightest in the vicinity of the constellation called Sagittarius and thus indicates the direction of the center of our galaxy. Figure I-9 will help you visualize our position in the Milky Way system. You can also see why our galaxy is called a spiral galaxy. The star clouds appear to outline spiral arms. It is very hard to realize how small our solar system is compared to our galaxy of stars which contains an estimated 200 billion stars, many of which are similar to our sun. Since we are on the inside looking out and around us, we cannot visualize very easily the outline of our galaxy. Astronomers orient our galaxy in space by referring it to directions of constellations. The two views in space (Figure I-9) require a total of six different constellations as direction indicators. In Chapter II, you will learn where these constellations are in the sky, with the exception of Argo, which can be seen only from the southern hemisphere.

Fortunately we can see and photograph with a telescope other galaxies far beyond the boundary of our Milky Way system, and astronomers have come to realize that many of them must be very similar in size and shape to our galaxy. In Chapter II you will find photographs of other galaxies which will remind you by their outlines of Figure I-9.

Now you are ready to *Point to the Stars*.

17

CHAPTER II

Pointing to the stars

There is a brotherhood among star gazers that knows no age or time. The lore of this brotherhood is not learned in one night. As you read this paragraph, feel urged to spend a few minutes each evening watching the stars. This is all that is required of you to be initiated into our fraternity. You will be a brother-in-the-stars and we promise you a lifetime of relaxation and continuous fascination.

Much of the enjoyment of star gazing comes from being able to locate and identify stellar objects which have been well-known through the ages. As a beginner, you might remark that this is difficult to do; the miniature sky pictured on a printed star chart, or map, has little apparent connection with the immense overhead. The sky has no lines or guides, no beginning and no end. How can you learn the different stars and where to find them in the sky?

Perceiving the constellations

The first thing you must do is to learn how to "see" the constellation group. This is not easy to do, especially if your eyesight is keen and you are away from city lights. An overwhelming number of stars are seen under these conditions. Looking up at the night sky, you are aware of more stars in the constellation region than are needed to form the constellation. How did the ancients make identifiable constellations stand out from a profusely star-studded background? They solved this problem by using only 500 brighter stars and ignoring the thousands of other fainter stars. This is what we meant by telling you that "seeing" the constellation is not easy. Happily, we can offer you a method which should help you find the constellation you are looking for.

It has been found that the human eye can be taught to discern an object hidden in a picture by directing the observer's attention to its outline. For example, can you see the head of an animal in the photograph (Figure II-1) on this page? Unless you have unusual perception, you will not be able to perceive a cow hidden in the picture. If you want a guiding clue, look at the outline in Figure II-2, on page 21. Then turn back to the original "hidden" picture. This figure which is hidden in its background (known as "camouflage") will stand out when your attention has been called to its outline. Once this happens, you will not only succeed in seeing the hidden object, but from then on you will never fail to find it when you look at the picture.

Let us apply this perceiving principle to identifying the constellations. Instead of picking out individual stars, concentrate on the *whole* configuration, using the lines on the map connecting the stars as a guide in the same way that you used the drawing on page 21 to find the cow in the picture on this page. Each time you see the picture in the future, you will recognize the cow without having to refer to the outline. In a similar way you will be able to learn the outlines of the constellations. Eventually, the constellations become so familiar to you that you do not need the map lines any more than you need to check constantly on height, color of eyes and other identifying features to recognize a friend. The constellations will be your friends forever, once you learn how to recognize them in the sky.

Pointing to the constellations

You have been given the basic principles for "seeing" the constellations, and the next step is for you to learn where to look for them in the sky. It will be of help if you will recall that your family or friends showed you where to look for your dog, or a strange bird, or an unusual sight—by *pointing* to it. If we were with you in person, we would show you where to look for the constellations by doing the same

Fig. II-1. Can you "see" the cow?

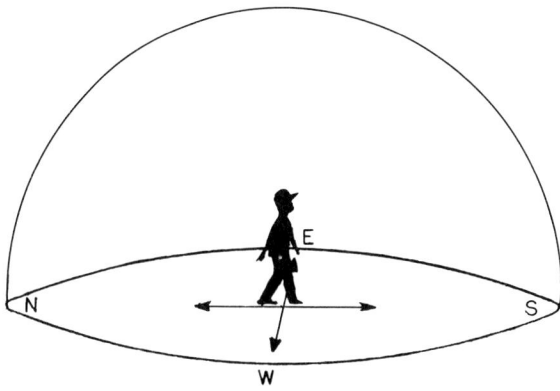

Fig. II-3. Little man faces south.

thing—pointing to them. Since this is not possible, we offer you our written directions in two steps which astronomers use. To understand these steps, think of how you would point out a bird in a tree to a friend. First you would have to *face* the direction of the tree. Then you would have to *point* directly at the bird. In astronomy, the first step, finding the direction, is called determining the *azimuth*. The second step, finding the height, is referred to as determining the sky height or *altitude*.

1. *Compass points—azimuth* Once you are able to point to, or "box"— as Boy and Girl Scouts and navigators call the step—the sixteen points of the compass at night, you will have mastered the most difficult part of pointing to the constellations (Figure II–3). Start with the south point, which may be learned by remembering that the daily noon position (Standard Time) of the sun indicates approximately south. Then, imagine yourself moving to *face* the horizon circle of 360 degrees, going from south to west to north to east back to south. Directions on the horizon circle are known as *azimuth points*.

2. *Sky height—altitude* The second half of the pointing method is to determine the sky height or altitude. Point your arm to the horizon and then move it straight up until you are pointing overhead. Your arm has just swept over a quarter circle (known as a *quadrant*), measured from the 0° point on the horizon to 90° overhead, the *zenith*. (You will have to tilt your head backward at an awkward angle to find the zenith.)

For the purpose of pointing to the constellations, this quadrant will be divided into four pointing positions—30°, 45°, 60°, 90° (Figure II–4). These divisions are sufficiently accurate for constellation identification.

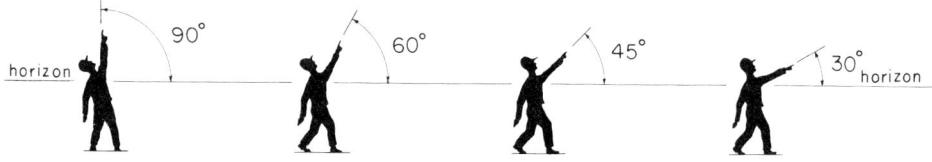

Fig. II-4.

We suggest that you practice with a program of pointing to the true south (and north, of course), to the other compass points, and to the sky heights or altitudes. This gives you a complete pointing system called *altazimuth coordinates*. Think of it as the *face and point* system.

Using the "little man" system to find a constellation

You will be helped by a "little man" shown pointing to *each* constellation at around 9 P.M., Standard Time, for the month printed at his feet. One of these months is underscored; the sky will look like the half-sky map on that page at 9 P.M. Remember, the little man is using the *face and point* system and so should you. The compass point tells in which direction to face and the angle of his arm tells how high to look.

If you are observing at 11 P.M., go to the next later month on the little man's compass. In the winter, if you are observing at 7 P.M., pick out the pointing directions given for one month earlier. In general, there will be about a half month's change for each hour's difference before or after 9 P.M.

You will recall from Chapter I that the stars near Polaris do not set but circle the pole and remain visible all the time. In case you do not find the name of a month printed under the little man, that constellation is not visible for that month at 9 P.M.

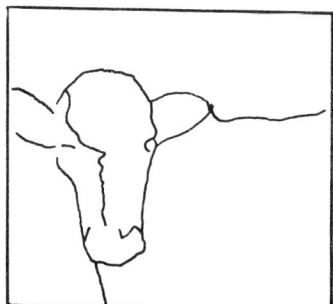

Fig. II-2.

21

Using the sky maps

When you have learned the *face and point* system (altazimuth coordinates) and can use it in your back yard, or wherever you may be, then you are ready to use the star maps in this chapter. Use these half-sky maps to learn what other constellations are visible and their positions in relation to the outlined constellation. You will find twenty-four maps, each of which covers almost half of the sky as seen at different times of the year. Some maps show the sky as it looks when you are facing south; the others show the sky as you face north. On each map, selected constellations are named. It is impossible to show the stars without some distortion in position when depicting a large area of the sky on a flat page. The stars at the top of the half-sky map may be actually overhead. These maps contain primarily the brighter stars, the ones which form the constellations. The stars are indicated according to diminishing brightness as follows: ✦ ● • • ·

On each map you will find the months with the corresponding hours below them to give you the right times to point to the night sky as depicted. For example, on the half-sky map for finding Orion (page 34), the southern sky will look like the map in February at 21 hours (9 P.M.) or November at 3 hours (3 A.M.), etc.

Two systems of naming the hours are used. The first is the common 12-hour system with which you are familiar. The other is based upon a 0 to 24-hour day. This numbering is used in astronomy, in the Navy and by some peoples in other parts of the world. Midnight is 0 hours, starting a new day. A.M. and P.M. are not necessary in this system. One o'clock P.M. becomes 13 hours, 6 P.M. is read as 18 hours, and 8 P.M. as 20 hours, and finally, 11:55 P.M. is read as 23 hours 55 minutes or simply $23^h 55^m$.

The constellations will be described one by one, and included in the description will be selected legends from mythology and folklore.* The lines connecting the stars emphasize the constellation described on the map page.

When a constellation is seen near the western horizon, it will appear upside down in comparison with the way it looks when it appears on the eastern horizon. Note the changing orientation of the Big Dipper and Orion in Figure 2, Chapter I.

The twinkling of the starlight you often notice is simply due to the agitated state of the many layers of our own atmosphere through

---

*There are as many versions of star stories as there were tribes and nations of ancient times. The legends we have used are the authors' choices.

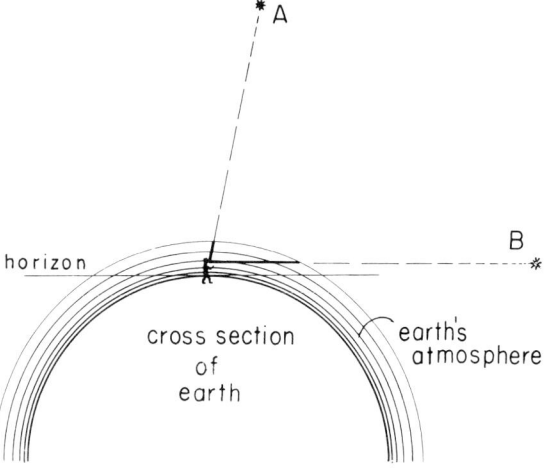

Fig. II-5. Best sky position for observing constellations. Our little man can observe stars better when they are high in the sky, at *A*, for example, than near the horizon near *B*. The light path through the earth's atmosphere is shorter for a star high in the sky than one near the horizon where the light is dimmed by absorption during its long path through the earth's atmosphere.

which the starlight has to travel. Also, you may notice that stars appear fainter near the horizon; some become so faint that you cannot see them. You can see from Figure II-5 that light from the stars near the horizon has a longer path through the earth's atmosphere. The longer the path, the more light the atmosphere absorbs. It is also the reason why you can often look directly at the sun near its setting or rising time, when it is seen as a red ball, far dimmer than when it is high in the sky.

Summary of directions for pointing to each constellation at 9 P.M. (21 hours)

  1. Turn to the page of the constellation you wish to find.
  2. Pick out the little man who is above the correct month of the year.
  3. Face the direction of the arrow.
  4. Raise your arm to the approximate height of the little man. With your extended arm in that position, you will be pointing to the constellation you want.

Example: Suppose you want to know where to look for Perseus at 9 P.M. (21 hours) at the beginning of March. Follow the steps shown above:

  1. Turn to page 74, where Perseus is described.
  2. Examine the compass to find where the little man is standing in March.
  3. Face the direction of the arrow, that is, northwest.
  4. Raise your arm to the height shown by the arm of the little man, that is, 30°.
  5. You will be pointing to about the center of the constellation of Perseus.
  6. Examine the guided outline on the half-sky map.
  7. Look up at the section of the sky where your arm is pointing.
  8. There is Perseus.

Detailed constellation charts

On the right-hand pages are given detailed charts of each constellation described. They show fainter stars and other objects which you may want to examine with binoculars or a small telescope. The stars are placed on the coordinate system generally used by astronomers.*

1. *Star names.* The brighter stars have individual names. Most of these were given to them by the Romans and the Arabs a long time ago, and described the places that stars mark in the figures. For instance, Rigel, in the foot of Orion, means the "foot" of the hunter.

2. *A simple way to mark the stars.* In 1603, Bayer, a German astronomer, made a famous set of star charts. To simplify the designations of the stars, he used Greek letters for the brighter stars of a constellation instead of their ancient names. He used Latin names for the constellations, and to show that a star was a member of a particular constellation he followed the genitive case rule of Latin. For instance, he named Sheraton, β Arietis (genitive case of Aries, meaning "of Aries"), which means that it is the second brightest star in this constellation because β is the second letter of the Greek alphabet. He designated the brightest star in each constellation α, the next β, and so on.

The Constellation Orion

Original Greco-Latin and Arabic names	*Simplified scientific names*
Betelgeuse (armpit)	α (alpha) Orionis
Rigel (foot)	β (beta) Orionis
Bellatrix (female warrior)	γ (gamma) Orionis
Mintaka (belt)	δ (delta) Orionis
Anilam (string of pearls)	ϵ (epsilon) Orionis
Alnitak (girdle)	ζ (zeta) Orionis
Saiph (sword)	κ (kappa) Orionis

(Unfortunately, Bayer was not consistent about this in all constellations.) The stars are indicated in this manner on the detailed constellation charts. The table will acquaint you with the names and appear-

*Right Ascension and Declination. The background coordinate system used for charting individual constellations is right ascension (R.A.) and declination (Dec.). These divisions will help to show you how you can estimate the relative sizes of constellations. Just as it is possible to locate any point on the earth by its longitude and latitude, so we can use a corresponding coordinate system on the celestial sphere to locate any stellar object. Right ascension is measured along the equator and is given in hours (symbol h); 24 hours complete the great circle which is the projection of the earth's equator. Declination (symbol δ) is measured (similar to latitude) in degrees north and south of the projected earth's equator; plus (+) degrees mean north δ and minus (−) degrees mean south δ.

ance of the Greek letters. When the letters of the alphabet are exhausted, numbers are used.

The Greek Alphabet

α alpha	η eta	ν nu	τ tau
β beta	θ theta	ξ xi	υ upsilon
γ gamma	ι iota	ο omicron	φ phi
δ delta	κ kappa	π pi	χ chi
ε epsilon	λ lambda	ρ rho	ψ psi
ζ zeta	μ mu	σ sigma	ω omega

3. *Magnitudes and colors.* The brightness of a celestial object is described by a number. The bright stars were called *first magnitude* by the ancients. The next class just fainter than these was said to be of the *second magnitude*. Still fainter ones were placed in the third, fourth, and fifth magnitudes. We still use this classification today, extended to zero and negative numbers for brighter objects and higher numbers for telescopic objects. For example, Sirius, the brightest star, is rated at magnitude -1.5 and Vega at $+0.1$. The companion to Sirius (see photograph on page 33) is rated at $+8.7$ magnitude. To indicate the stars of the different magnitudes, various distinctive symbols are used on the detailed charts. These are shown in the diagram (Figure II–6). Stars differ in color as well as in brightness. When you see white or blue-white stars, you are observing stars with very hot surfaces. Orange and red stars do not have such high surface temperatures.

4. *Star distances.* The stars are so far away that it is not convenient to tell their distances in miles. A longer measuring stick is used and is called a *light-year*. This is the distance that light travels in one year, and is nearly 6,000,000,000,000 miles. The nearest star, α Centauri, invisible from the north middle latitudes, is 4.3 light-years distant.

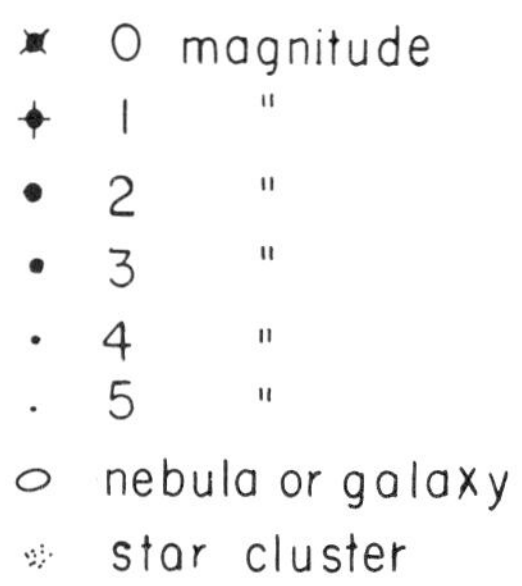

Fig. II-6. Key to detailed constellation charts.

✭ 0 magnitude
✦ 1 "
• 2 "
• 3 "
· 4 "
. 5 "
○ nebula or galaxy
✲ star cluster

5. *Double stars.* To the naked eye, some stars appear as a single star, but when examined through binoculars are revealed as two stars. Often these two stars belong to one another; that is, they revolve about each other as do the planets about our sun. In fact, they may be separated by distances comparable to those between the planets and the sun.

In the following pages we will refer to the *apparent* separation of double stars, which is given in seconds, or minutes, of arc to indicate the measurement of an angle. There are 60 seconds (60″) in a minute of arc; 60 minutes (60′) in one degree (1°). You can get an idea of how much a degree of arc is by knowing that the diameter of the full moon is about one-half degree. The pointers to the Pole Star are 5 degrees apart (Figure II-7).

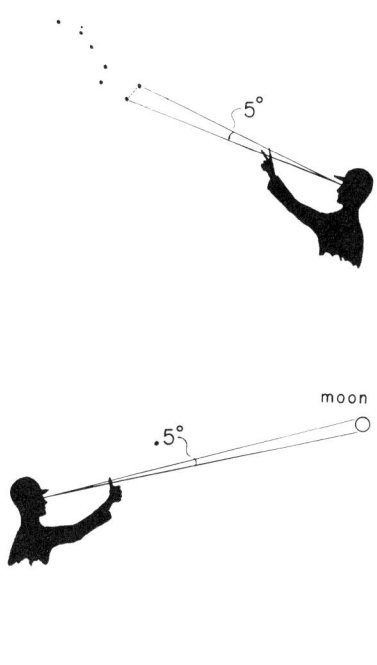

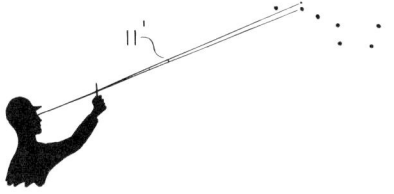

Fig. II-7. Angular separation of the pointers to the Pole Star: 5°. Angular diameter of the moon: .5°. Angular separation of Mizar and Alcor: 11′.

Some of the double stars referred to in the individual constellations appear closer together than 3 minutes of arc, and require binoculars to note that they are double. Closer pairs require a telescope to observe each component separately.

First use of binoculars

When you are ready to look for the fainter objects described in the sections, *With Binoculars,* these pointers may help you. Study the detailed constellation charts and determine the naked-eye star closest to the faint object you wish to find. Face, point and identify the constellation, then look for the naked-eye star. Concentrate your attention by gazing intently on this star and, without taking your eyes off it, raise your binoculars to your eyes. The naked-eye star which is your guide to the fainter object will be near the center of the "field" of the binoculars. Shift your binoculars slightly in the direction of the faint object that you are looking for.

First use of the telescope

Directions for the use of a telescope in astronomy are usually supplied with the telescope. In case you do not have any instructions, we suggest the following: Sight along the tube and aim it in the proper face-and-point direction to the naked-eye object closest to the object you are looking for. Next, look through the eye-piece to center the desired object. Telescopes which are for astronomical use show objects upside down and reversed from the way we see them with the naked eye. You can test this reversal by looking at the moon when it is not full; look first with the naked eye, then with the telescope. It will take you a little while to get used to this reversal, but, as you know, from the space point of view there is no distinction between right side up and upside down.

Glimpses of the universe

The photographs and drawings at the lower right corner of the detailed map page show a variety of astronomical objects and concepts basic to your understanding of the universe. Many of the objects shown are numbered according to Messier's catalogue. This French amateur astronomer was a "comet hunter" and made this famous list of 103 objects that looked to him like comets. The "comets" turned out to be bright masses of gas, great clusters of stars, or galaxies.

TAURUS

"Canst thou find the cluster of the Pleiades?"
  Book of Job

Month	Feb		Jan		Dec	Nov		Oct		Sept		Aug
Hour (S.T.)	17h 5pm	18h 6pm	19h 7pm	20h 8pm	**21h** **9pm**	22h 10pm	23h 11pm	24h 12pm	1h 1am	2h 2am	3h 3am	4h 4am	5h 5am

1. Face South.
2. Map shows the sky for the months listed and the times given below the months.

1. Select the month.
2. Face the compass point.
3. Point your arm as high as the Little Man's.

THE BULL

In ancient Phoenicia, the beautiful maiden Europa, daughter of Inachus, was fond of the animals in her father's herds. Jupiter became enamored of her, but realized that her father would never permit him to visit their home. Jupiter learned of Europa's fondness for animals and concocted a scheme to carry her away in spite of the close watch Inachus kept on her. He took the form of a snow-white bull and mingled with the other cattle, which he knew Europa would visit each day. Europa could not help noticing this beautiful bull because he did not snort with anger nor paw the ground in rage. He even permitted her to pet him. Encouraged by the animal's gentleness,

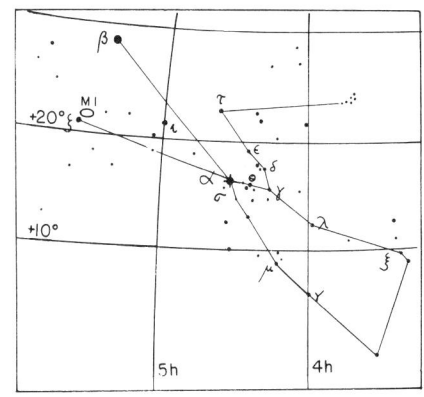

Europa tried to ride on his back for short distances in the corral. During one of these rides, Jupiter galloped off so quickly that she had to hang on tightly or fall. The disguised god thereupon set out for the island of Crete with his prize captive finding it impossible to escape because the bull was swimming in the deep ocean. Do you see why only the upper half of the body of Taurus is shown in the constellation figure?

For some unknown reason, the maiden Europa was not brought into the sky picture, but two groups of sisters, the Pleiades and the Hyades, are depicted with TAURUS as their protector against the wiles of Orion (see page 34).

With the Naked Eye. α Tauri, Aldebaran, means "the Follower." It is about 70 light-years away and is red in color. The Hyades is a V-shaped group of stars beginning near Aldebaran, forming the face of the bull. These stars form a cluster moving through the sky, and are about 120 light-years away. The Pleiades, a compact little dipper, is another distinct cluster called an "open cluster." Some observers see only six, others up to ten stars.

With Binoculars. El Nath, "the Tip," β Tauri, is a beautiful white star and is less than 200 light-years away. The Hyades and the Pleiades are beautiful fields of stars. How many stars can you count in the Pleiades?

With the Telescope. Look for the same objects you can see with the binoculars. Look at M 1, the Crab nebula, which is near ζ Tauri.

CRAB NEBULA — MESSIER 1

Star gazers in 1054 A.D. were astounded to see a bright star suddenly appear in the constellation Taurus. The star gradually became fainter according to records of this *nova*, or new star. This photograph taken in the position where the nova appeared shows what astronomers believe to be the results of a great stellar explosion seen 900 years ago. We still observe hot gases rushing away from the center of the cloud-like formation. Recently, astronomers believe that they have identified the remnants of the exploded star; it is called a *pulsar* because the light from it brightens, or "pulses," about 30 times a second. (200-inch photograph, Mt. Wilson and Palomar Observatories.)

CASSIOPEIA

*"... heaven troubled Queen with scanty stars
But lustrous in the full-mooned night sits the Queen ...
Uplifted hands, seems wailing for her child."*
PHENOMENA, by Aratus of Soli

Month	Feb		Jan		**Dec**	Nov		Oct		Sept		Aug
Hour	17h	18h	19h	20h	**21h**	22h	23h	24h	1h	2h	3h	4h	5h
(S.T.)	5pm	6pm	7pm	8pm	**9pm**	10pm	11pm	12pm	1am	2am	3am	4am	5am

⇧
1. Face North.
2. Map shows the sky for the months listed and the times given below the months.

◁
1. Select the month.
2. Face the compass point.
3. Point your arm as high as the Little Man's.

THE QUEEN ON THE THRONE

A queen, as everyone knows, must always maintain dignity on the throne, so we cannot help feeling sorry for Queen Cassiopeia, wife of Cepheus. She and her daughter Andromeda were beautiful, and the Queen Mother could not resist the temptation to boast of this beauty. Sea nymphs, the goddesses of the sea, were offended when mortal Cassiopeia dared to praise herself as being more beautiful than they. The sea nymphs appealed to the heavenly council, which voted to punish Cassiopeia by embarrassing her. The Queen on the Throne was changed into stars in the northern sky. As she revolved

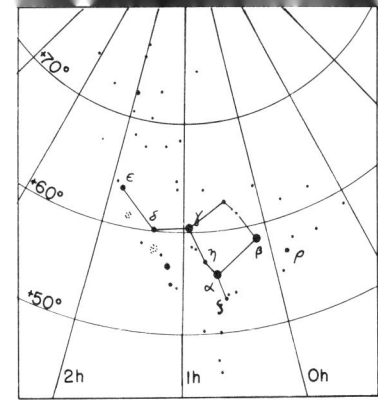

around Polaris, she would, for six months of the year, be sitting on her throne head down, like a tumbler. This position would shame her and remind her and other mortals never to be so audacious as to compare themselves to the immortal celestial ones. Then, to show the justice of the council, Queen Cassiopeia was permitted to be upright on her throne, as a dignified queen should be, for the other six months of the year.

Is Cassiopeia now in an unqueenly position, or has she been restored to her regal dignity?

The most conspicuous figure in CASSIOPEIA is W- or M-shaped. Cassiopeia is just across the pole from the Great Dipper. Both are circumpolar constellations; when one is high in the sky the other is low over the north point.

With the Naked Eye. Trace out the W. Find α Cassiopeiae, Schedar. This star is slightly variable, and it is about 200 light-years away. For the last quarter century, γ Cassiopeiae has remained constant in brightness at 2.6 magnitude. It now appears to be brightening; by the end of 1966 it had increased to 2.2, so that it bears watching in the future. For comparison, β Cassiopeiae is constant at 2.28 magnitude.

With Binoculars. Near ζ Cassiopeiae is a little circlet that looks like a small crown. Study the beautiful star sights near δ Cassiopeiae.

With the Telescope. η Cassiopeiae is double with magnitudes 3 and 7; 10" separation. Herschel discovered this double in 1779. α Cassiopeiae is double: magnitudes 2 and 9; 1' apart. Look for the clusters marked on the diagram.

A NEARBY GALAXY; NEW GENERAL CATALOGUE NO. 147

A galaxy consists of a great number of stars concentrated in one region of space. Often gas and dust are found between the stars. Galaxies are sometimes called island universes because of the great empty regions which separate them. Astronomers believe that there are a billion galaxies within the reach of our present telescopes. The most distant ones that have been observed may be as far as 10 billion light-years away. The galaxy shown here is within 1,800,000 light-years of our galaxy, and therefore it is considered "nearby". As you look at the photographs of other galaxies, notice how they vary in shape. (200-inch photograph, Mt. Wilson and Palomar Observatories.)

CANIS MAJOR / CANIS MINOR

*"Hail, mighty Sirius, monarch of the suns!
May we in this poor planet speak with thee?"*
THE STARS, by Lydia Huntly Sigourney

Month	Mar		**Feb**		Jan		Dec		Nov		Oct
Hour	19h	20h	**21h**	22h	23h	24h	1h	2h	3h	4h	5h
(S.T.)	7pm	8pm	**9pm**	10pm	11pm	12pm	1am	2am	3am	4am	5am

1. Face South.
2. Map shows the sky for the months listed and the times given below the months.

1. Select the month.
2. Face the compass point.
3. Point your arm as high as the Little Man's.

THE GREAT DOG / THE LITTLE DOG

When Procris, daughter of Thespius, married Cephalus in ancient Greece, she received two unique wedding presents. One gift was a dart, which, once hurled, would seek out its target, trail it and then make a hit. The other present was the hound dog Laelaps, which, once started on a scent left by an animal, would always get his quarry. Procris valued these gifts highly. Her husband Cephalus persuaded her to let him use them on a hunting trip. While he was gone, Procris became lonely for him, as she loved her husband very much, and decided to join him at his camp. Wishing to surprise him, she left secretly. At the camp, watchdog Laelaps growled an alert to Cephalus, who thought that the rustle made by Procris was that of a wild animal. He quickly hurled the unerring dart into the dark, silencing Procris forever.

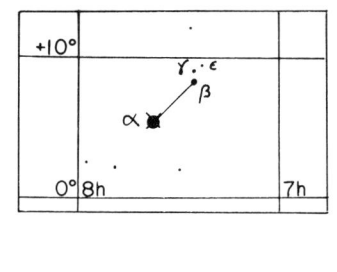

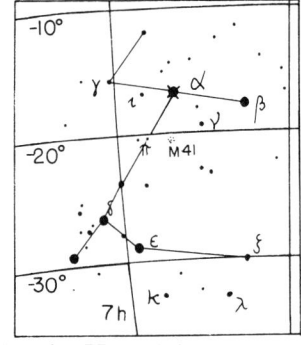

Laelaps was placed in the skies following Orion, where he is doomed forever to chase Lepus the hare in the southern sky just out of reach of his jaws. Laelaps was permitted, however, to be the weather watchdog for the Egyptians across the Mediterranean. From his orbiting position high in the sky, he could watch the Upper Nile River. When the Nile floodwaters started north, he would alert the people to have their fields ready for irrigation. How did he do this? The priests in the temples would daily watch the rising sun. As soon as Canis Major showed his jaws, as represented by Sirius rising shortly before the sun did, the priests would send the signal over the countryside. Would today's inventors of target-seeking rocket missiles be surprised to learn that the idea of such weapons was first thought of thousands of years ago?

CANIS MAJOR is noted because it contains the brightest star in the whole sky, Sirius, whose magnitude is -1.5. It is commonly known as the Dog Star.

With the Naked Eye. Find Murzim, β Canis Majoris, and Sirius, α Canis Majoris. Sirius is one of the nearest stars, only 8.7 light-years away, and is 23 times as bright as the sun if both are viewed from the same distance. Procyon, α Canis Minoris, is one of the brightest stars in the sky. It is yellow, and has a magnitude of 0.5. It is 11 light-years away and is six times as bright as the sun.

With Binoculars. Study the color of Sirius. If Sirius were not so brilliant you could see its companion in a small telescope. ν Canis Majoris is a triple star.

With the Telescope. Study ν Canis Majoris and the cluster M 41. ϵ Canis Majoris is a double star; magnitudes 2 and 9; separation 7".

SIRIUS AND ITS COMPANION

Sirius is the brightest star in the sky and is only 8.7 light-years away. Note the faint companion, only 1/10,000 times as bright and, therefore, very hard to detect. In this photograph a hexagonal opening was used in front of the telescope to give the image of bright Sirius this shape, which allows the companion star to be seen.

The companion is a white dwarf star not much larger than the earth, with its surface hotter than the sun's and a density so great that a sample of its contents the size of a golf ball would weigh several tons. (24-inch refractor photograph, Sproul Observatory.)

ORION

*". . . behold the Hunter
With stretched arms almost fathoming the skies."*
　　　　　THE SPHERE, by Marcus Manilius

Month	Mar		**Feb**		Jan		Dec		Nov		Oct
Hour	19h	20h	**21h**	22h	23h	24h	1h	2h	3h	4h	5h
(S.T.)	7pm	8pm	**9pm**	10pm	11pm	12pm	1am	2am	3am	4am	5am

1. Face South.
2. Map shows the sky for the months listed and the times given below the months.

1. Select the month.
2. Face the compass point.
3. Point your arm as high as the Little Man's.

THE HUNTER

Orion was an important hunter-warrior. In return for ridding the kingdom of Chios of the beasts attacking the people, Orion had been promised beautiful Merope for his bride by her father, King Enopion. When the king refused to keep his promise, Orion tried to take Merope by force, but Enopion became enraged and had Orion blinded. Unable to hunt further, Orion was wandering sightless over the earth when the god Vulcan sent the Cyclop Cedalion, his one-eyed assistant, to turn Orion about so he was facing the rising sun. This restored his sight. Orion went back to his hunting and met Diana, goddess of the hunt. Diana's brother, Apollo, became alarmed that his sister might break her vow never to marry and worried that there might be a ro-

34

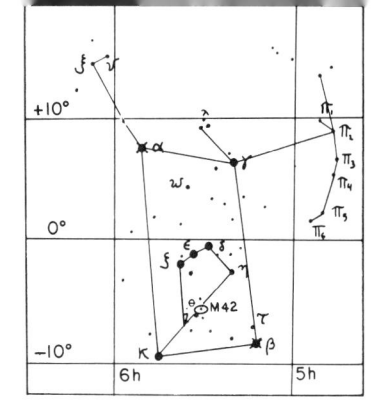

mance between her and Orion. To prevent this, he tricked Diana into a target shooting test, which was to hit a small dark object far out in the sea. Of course Diana did not miss, but the target was the head of Orion. Grieving at what she had done, Diana made Orion into a constellation in the sky with his hunting dogs, Canis Major and Canis Minor, close by. Who do you think were Orion's sky neighbors? The seven daughters of Pleione, the Pleiades, and also the seven daughters of Aethra, the Hyades! Jealous Jupiter, in the guise of Taurus (see page 28), forever keeps Orion separated from the lovely maidens.

ORION is the brightest of all the constellations.

With the Naked Eye. The three stars forming Orion's Belt make a striking figure. The star above is Betelgeuse, α Orionis. The blue-white star below the Belt is Rigel, β Orionis, which means "foot." The curved line of stars represents the lion's skin over Orion's arm. The Belt is just 3 degrees in length and is sometimes called the Yardstick.

With Binoculars. α Orionis, Betelgeuse, is a variable star with a range of 1 magnitude. It is about 400 times as large as the sun and approximately 300 light-years away. β Orionis, Rigel, is about 540 light-years away and approximately 18,000 times as bright as the sun. In the Sword is the Great Nebula in Orion. This great mass of glowing gas is M 42.

With the Telescope. δ Orionis is double; magnitudes 2 and 7. In the Great Nebula, notice the Trapesium—four stars close together. β Orionis, Rigel, is also a double star; it is difficult to see the companion star as it is 7 magnitudes fainter than Rigel.

"HORSEHEAD" NEBULA IN ORION

This unusual-looking nebula is south of ζ Orionis. Here you see a dark cool gaseous nebula outlined against a background of luminous gas. The dark nebula absorbs so much starlight that you cannot see the stars behind it. The stars you see may be in the foreground or they may be imbedded in the bright gas. A typical dark nebula may be 500 light-years across. (200-inch photograph, Mt. Wilson and Palomar Observatories.)

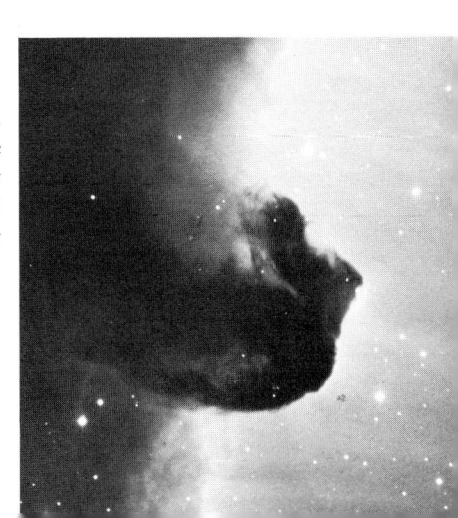

URSA MAJOR

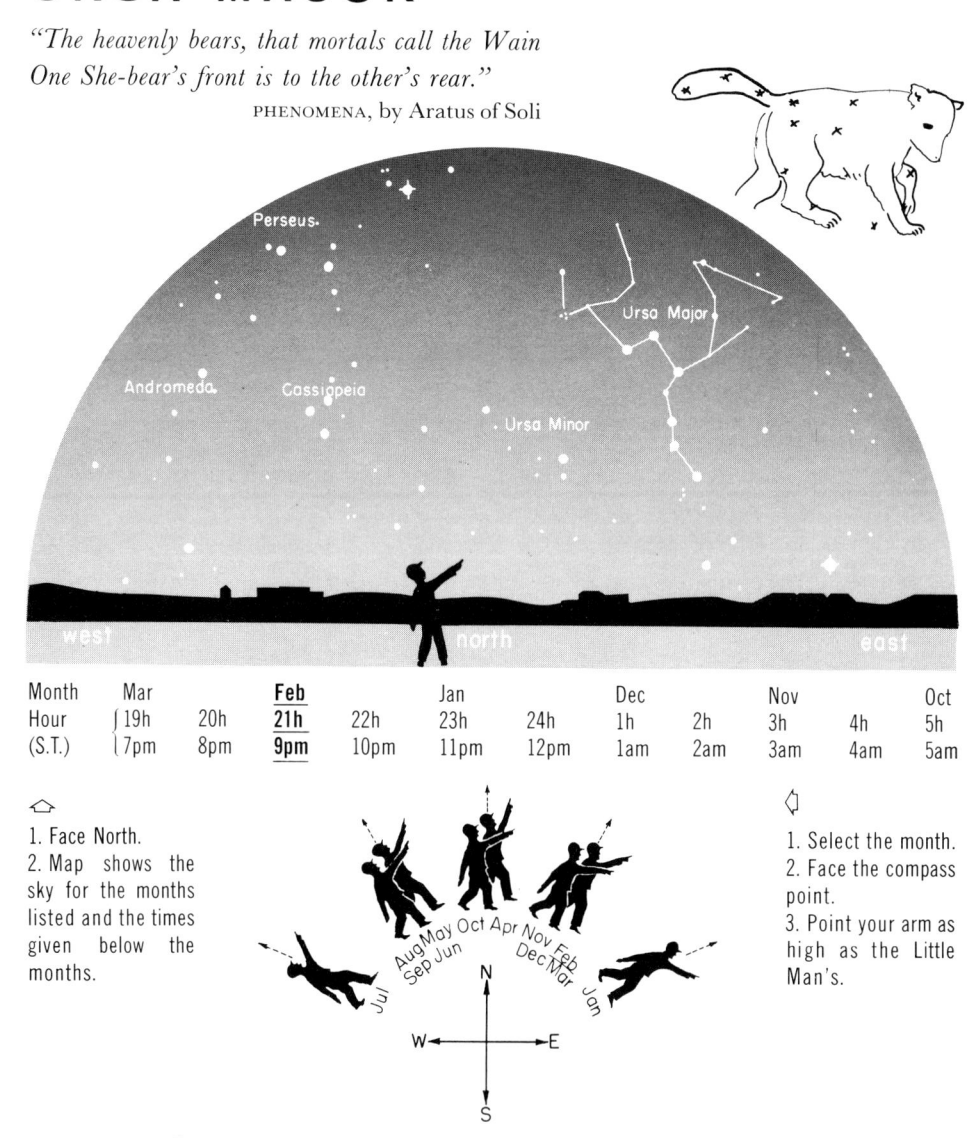

*"The heavenly bears, that mortals call the Wain
One She-bear's front is to the other's rear."*
       PHENOMENA, by Aratus of Soli

Month	Mar			**Feb**		Jan		Dec		Nov		Oct
Hour	19h	20h		**21h**	22h	23h	24h	1h	2h	3h	4h	5h
(S.T.)	7pm	8pm		**9pm**	10pm	11pm	12pm	1am	2am	3am	4am	5am

1. Face North.
2. Map shows the sky for the months listed and the times given below the months.

1. Select the month.
2. Face the compass point.
3. Point your arm as high as the Little Man's.

THE GREAT BEAR

One of the myths ancient Greeks loved to relate concerned the plight of two mortals, beautiful Callisto and her son Arcas, who incurred the wrath of jealous Juno, the goddess queen and wife of Jupiter, King of Olympus. Callisto was changed into a bear and separated from Arcas, who was left an orphan boy. Being separated was tragic enough for Callisto and Arcas, but imagine the horror of the situation when grown-up Arcas, now a hunter, came across a docile bear in the woods near his home. Callisto could not communicate with her son, and Arcas did not know the tame bear was his mother!

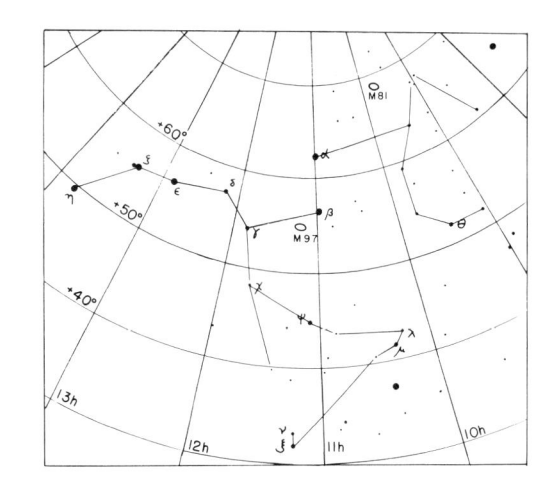

He was about to let fly an arrow to kill the bear when Jupiter, acting in his role of judge, caused the son, too, to be changed into a bear. To illustrate forever his humanitarian character, Jupiter also proclaimed that Mother Bear, Callisto, and Son Bear, Arcas, be changed into stars and placed in the northern sky. Jupiter lifted the bears by their tails. The bears were heavy and it was a long distance from the earth to the sky; their tails stretched from stubby to long ones. Can you see why early Teutonic tribes saw these constellations as a wagon, or chariot, called a Wain?

URSA MAJOR is famous as being the home of the Big Dipper, the best-known group of stars in the northern hemisphere. Most of the stars in the Big Dipper are about the same distance from us and form a large group of stars moving together in space.

With the Naked Eye. The stars at the end of the bowl are the Pointers, and a line through them leads you to the Pole Star. If you look carefully at ζ Ursae Majoris, Mizar, you can see an extra star, Alcor. Being able to see faint Alcor was once a test of eyesight for army archers. Mizar is about second magnitude and Alcor, fourth. They are about 70 light-years away from us.

With Binoculars. Study the pairs of stars, ν and ζ, that make the feet of the bear, and you will see beautiful fields of faint stars. Look at Alcor and Mizar and notice how far apart they seem. They are 11' apart. Between Alcor and Mizar is a faint star that was mistaken for a solar planet 200 years ago (what could a planet be doing here?) and named Sidus Ludovicianum.

With the Telescope. Study the field near γ Ursae Majoris. Mizar itself is a double. The components are of magnitudes 2 and 4, separation 14".

MESSIER 81, SPIRAL GALAXY

Messier 81 is a distant star system and perhaps something like our own Milky Way system. There are so many stars near its center that they blend together to make a great patch of light on the photograph. M 81 will appear as a minute fuzzy patch through a small telescope. (200-inch photograph, Mt. Wilson and Palomar Observatories.)

GEMINI / CANCER

"Tender Gemini in strict embrace
Stand clos'd and smiling in each other's face."
          by Marcus Manilius

Month	May		Apr	Mar		Feb		Jan		Dec
Hour		20h	**21h**	22h	23h	24h	1h	2h	3h	4h	5h
(S.T.)		8pm	**9pm**	10pm	11pm	12pm	1am	2am	3am	4am	5am

1. Face South.
2. Map shows the sky for the months listed and the times given below the months.

1. Select the month.
2. Face the compass point.
3. Point your arm as high as the Little Man's.

THE TWINS / THE CRAB

The twin brothers, sons of the god Jupiter and the mortal Leda, were not identical twins. Pollux was immortal, while Castor was born mortal like his mother. The twins hunted side by side and fought in wars the same way. Castor, the mortal twin, was killed in battle, while Pollux survived the combat. Realizing that he would be separated from his brother, Pollux in his grief implored Jupiter to rescind his immortality so that he could join his twin brother in death. Their father was so moved by this brotherly love that he restored life to Castor on the condition that immortal Pollux would spend half his time in Hades, the underworld. Pollux agreed to do so. To reward this good deed, the forehead of each brother was brightened with a star and the twins were equally immortalized as a single constellation. Is Pollux still keeping his promise about spending half his time in Hades?

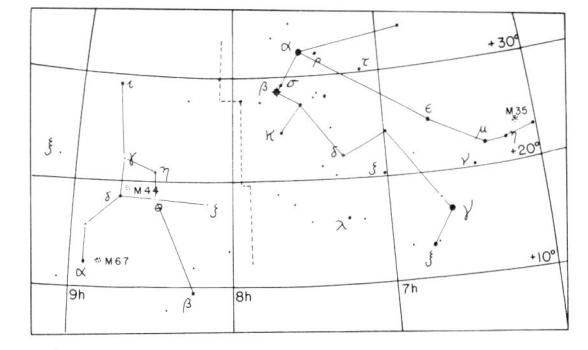

Cancer, although a crab, was a secret agent of Juno, goddess wife of Jupiter and enemy to Hercules. When the huge Kneeler (see page 56) was in the marshes battling Hydra, the many-headed water snake, Cancer, on orders from Juno, pinched his toes in order to distract him. When Hercules crushed Cancer, Juno had the crab placed in the skies as a reward.

GEMINI is marked principally by Castor and Pollux, and CANCER has no bright stars.

With the Naked Eye. α Geminorum, Castor, the northern star of the twins, is a white multiple star of 1.6 magnitude, and is about 45 light-years away. Pollux, β Geminorum, the southern twin, is of magnitude 1.2, is yellow and is 40 light-years away. Pollux is designated β, although it is brighter than Castor. Between δ Cancri and η Cancri is a faint hazy spot—a cluster of many stars called Praesepe. Sometimes it is called the Beehive, or Manger, marked M 44 on the map.

With Binoculars. Near η Geminorum is a compact cluster, M 35, that you can just see as a tiny speck. Search the lower part of the constellation near the Milky Way for beautiful star fields. Study Praesepe, M 44, and notice how many stars you see in this famous group.

With the Telescope. Notice the components of Castor: magnitudes 2 and 3; 4″ apart. They revolve around each other in a period of 380 years. At a separation of 1′ you find a distant companion to this system which is of the ninth magnitude and is very red. You can probably count over 300 stars in the Beehive. Study the cluster which is near α Cancri. Herschel found over 200 stars in it. How many can you see? This is M 67, which is probably one of the oldest galactic clusters that we know of.

CASTOR AS A DOUBLE STAR

Through a small telescope you will see Castor as two stars, differing by one magnitude, with the fainter, known as a companion, appearing to the right and above in an inverting telescope. It has been found that the companion revolves around the brighter star during a period of 380 years, as shown in the diagram. Components of double star systems revolve around each other in the same manner in which the earth revolves around the sun.

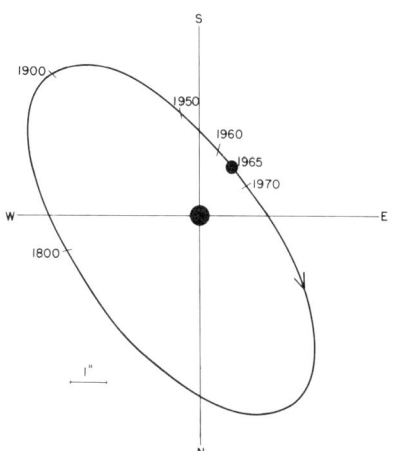

LEO

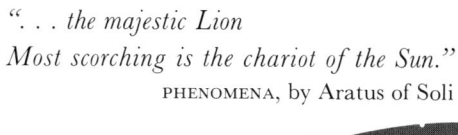

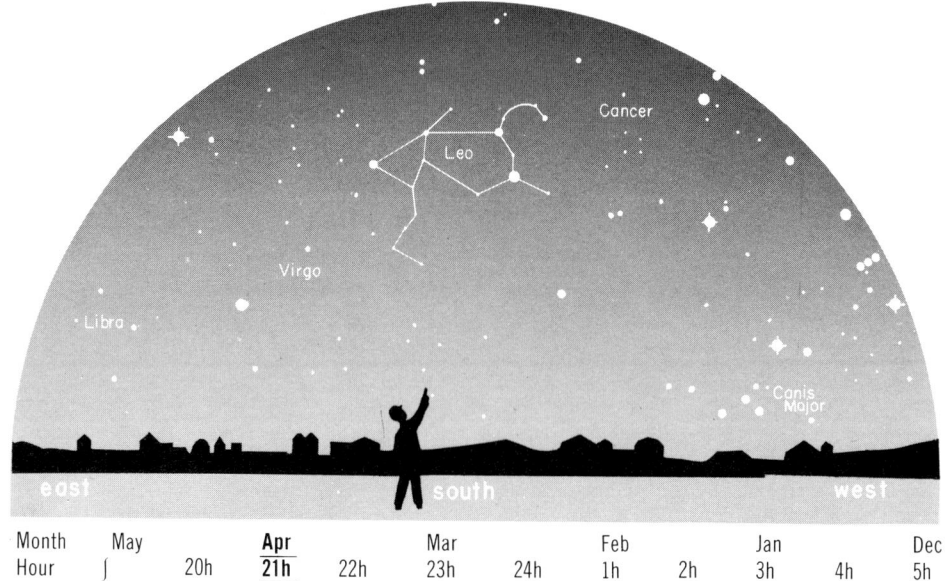

*"... the majestic Lion
Most scorching is the chariot of the Sun."*
              PHENOMENA, by Aratus of Soli

Month	May		**Apr**	Mar	Feb		Jan		Dec
Hour (S.T.)	{	20h	**21h**	22h	23h	24h	1h	2h	3h	4h	5h
		8pm	**9pm**	10pm	11pm	12pm	1am	2am	3am	4am	5am

1. Face South.
2. Map shows the sky for the months listed and the times given below the months.

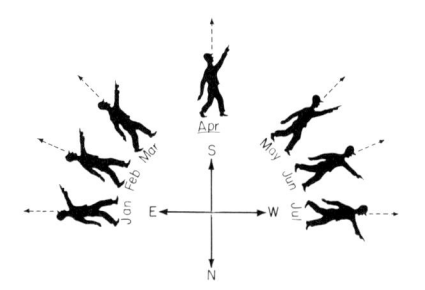

1. Select the month.
2. Face the compass point.
3. Point your arm as high as the Little Man's.

THE LION

Leo, a majestic lion, was sent down from the moon to roam the Nemean forests near Cleonae, a town in ancient Greece. Known as the Nemean lion, Leo was noted for a skin so tough and strong that no mortal arrow could penetrate it. He was regarded not only as the King of the Beasts, but was also master of forest and field because, of course, no hunter could slay him. His security seemed assured until strongman Hercules was ordered—in the first of his twelve labors (see Hercules, page 56)—to bring back Leo's hide. Leo naturally resisted, and at first the lion seemed to be winning because all the arrows shot at him bounced off his skin. Hercules then pulled up a young oak tree by the roots and struck Leo with it, stunning him. He was then able to crush the lion in his massive arms. Out of Leo's tough skin, Hercules made a kind of battle armor which he wore thenceforth. Since Leo was of heavenly origin, the gods returned him to the heavens—not on the moon again—but as an honored star group bearing his name. Could it be Leo's Nemean body that is so apparent in the Androsphinx sculpture and in the inscriptions on the wall of the Temple of Rameses at Thebes in Egypt?

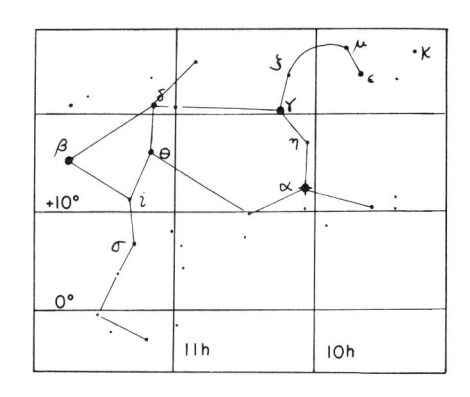

LEO consists of the "sickle" at the western end of the constellation, forming the head and forepart of the lion, and a triangle marking the hindquarters.

With the Naked Eye. Regulus, α Leonis, is the bright star at the end of the handle of the sickle. With Antares, Aldebaran, and Fomalhaut, it is one of the four Royal Stars of Persia. Sometimes it is called the Lion's Heart. Denebola, β Leonis, is the tip of the Lion's tail.

With Binoculars. α Leonis is a double star.

With the Telescope. γ Leonis is among the finest northern sky doubles; magnitudes 2 and 4; separation 4″. α Leonis has a deep blue companion; magnitude 8; separation, 3′.

A NAKED-EYE COMET—HALLEY'S

Halley's comet is one of the very few which can ever be seen with the naked eye. This photograph was taken at its last appearance in 1910. Halley's comet is scheduled to make its next appearance around 1984, when it is again near the sun. Comets revolve around the sun in the same manner as do planets, but their orbital paths generally take them beyond the earth's orbit. The stars are short streaks on the photograph because Halley's comet was moving rapidly across the sky and the telescopic camera followed the comet. (Yerkes Observatory.)

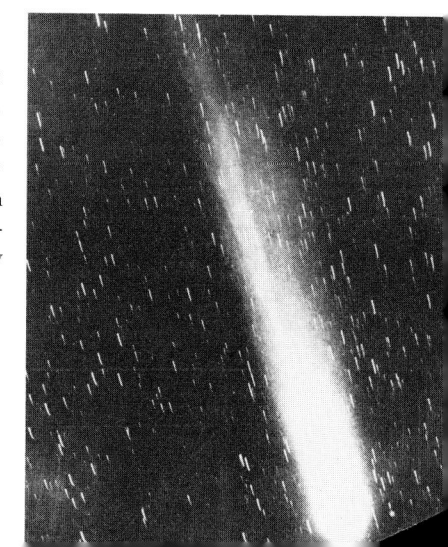

AURIGA

*"The charioteer, who gained by skill of old
Heaven and his name; as first four steeds he drove
On flying wheels seen."*
　　　　　　　THE SPHERE, by Marcus Manilius

Month	May		Apr		Mar		Feb		Jan		Dec
Hour		20h	**21h**	22h	23h	24h	1h	2h	3h	4h	5h
(S.T.)		8pm	**9pm**	10pm	11pm	12pm	1am	2am	3am	4am	5am

1. Face North.
2. Map shows the sky for the months listed and the times given below the months.

1. Select the month.
2. Face the compass point.
3. Point your arm as high as the Little Man's.

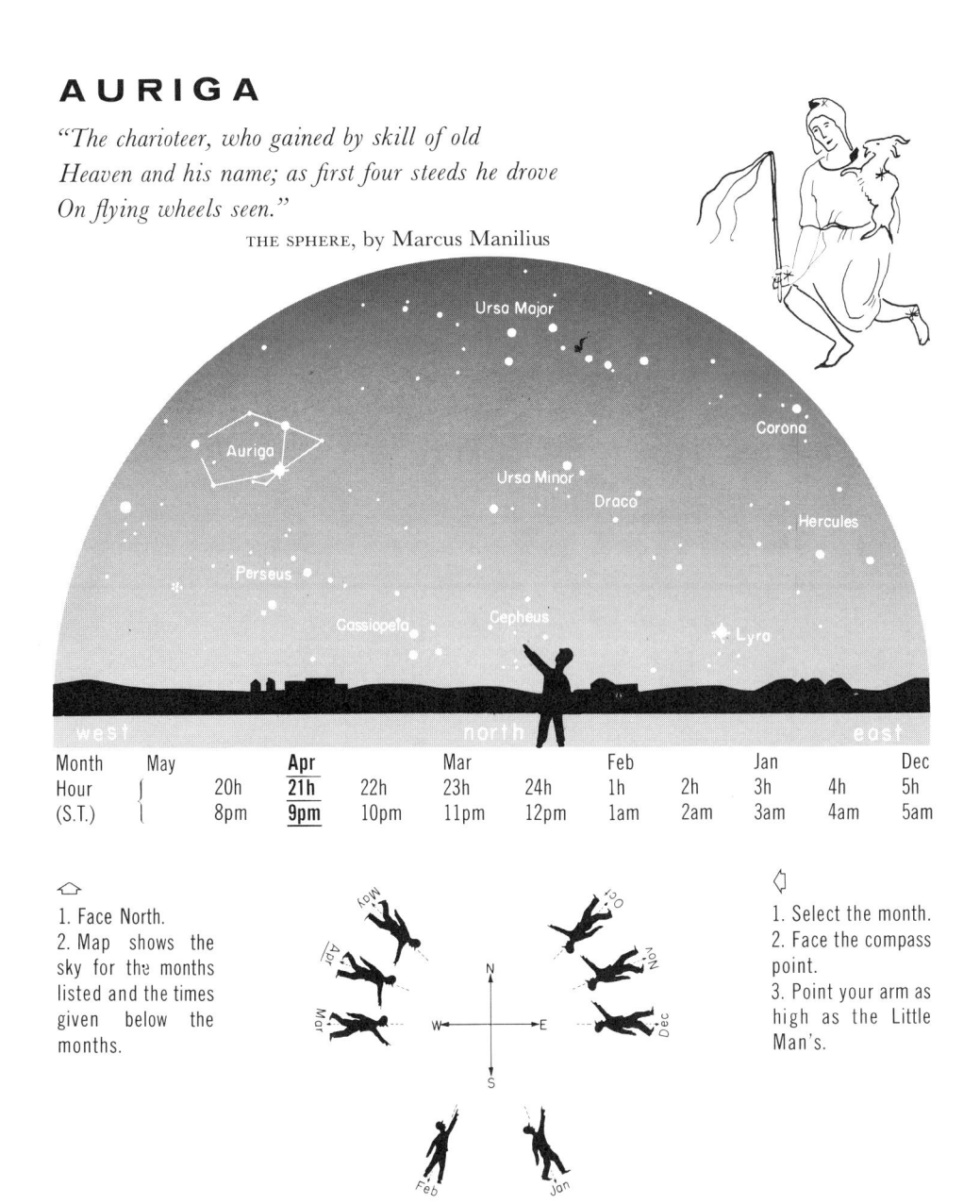

THE CHARIOTEER

Auriga was the crippled son of the goddess Minerva and the blacksmith god, Vulcan, who was lame, also. Courageously, Auriga set about to make his transportation easier by inventing a smooth-riding, four-horse chariot. Jupiter and the other gods thought so much of this improved travel vehicle that they honored Auriga by placing him in the sky with a wagon whip in his right hand. How Auriga came to be holding a nanny goat on his left arm we are not certain. But ancient stories relate that the daughters of the Cretan king Melis-

42

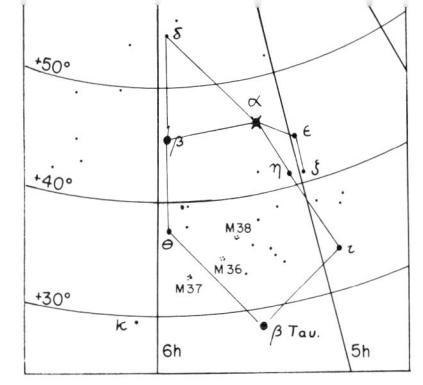

seus were assigned to feed the infant Jupiter, and that the baby-food formula they gave him was a mixture of milk from Capella, the goat, and honey. It may be that Jupiter wanted to remember his babyhood and trusted only gentle Auriga to carry the little she-goat, Capella, everywhere he went. It would be easy to say that Auriga was the first milkman, wouldn't it?

AURIGA lies along the Milky Way. The name of Capella, α Aurigae, the bright star in this group, means the Goat, and near it is a group of three stars called the Kids. The stars (if we borrow one from Taurus) are arranged in a pentagon, a five-sided figure.

With the Naked Eye. Capella has the same color as our sun. It is 42 light-years away and about 150 times as brilliant as the sun. At about 9 o'clock during the middle of January, Capella is almost directly overhead if you live at 40° north latitude. Find the three stars that represent the Kids. The star farthest south in the pentagon, the bright star at the bottom, is El Nath, β Tauri, just borrowed for the occasion.

With Binoculars. All over the area of Auriga you will find many good fields of stars, since the Milky Way is rich here. Look for clusters marked in the diagram. Study the color of Capella. It is magnitude 0.2, that is, nearly a magnitude brighter than first. This makes Capella fifth in order of brightness of all the stars in the sky.

With the Telescope. Study more carefully the clusters noted above and look for many doubles within and just around the edges of the pentagon.

THE SUN

Our sun is a star and all stars are suns. Our sun is the only star which we can observe as a disk. All others are so far away they appear only as points of light. In this photograph you can see several sunspot groups representing large disrupted regions. They appear dark only because they are some 2000° cooler than the rest of the surface, which is about 6000° centigrade. All stars are self-luminous and are gaseous throughout, with temperatures on the order of millions of degrees at the center. The main constituents of stars are hydrogen and helium, with a trace of the majority of the elements that we know on earth. The vast amount of radiation which pours out over millions of years is produced by the conversion of hydrogen into helium. (Mt. Wilson and Palomar Observatories photograph.)

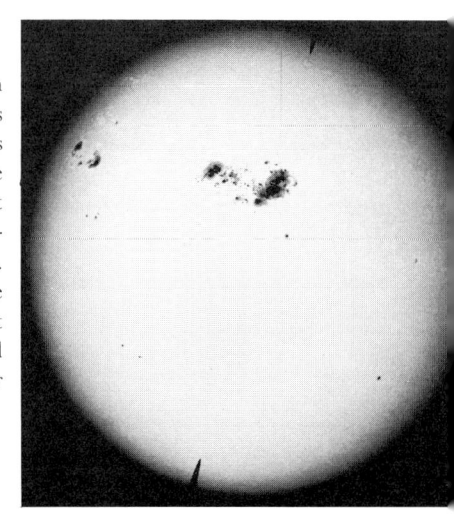

URSA MINOR

*"The name of one is Cynosura
. . . guides Phoenicians over the main."*
PHENOMENA, by Aratus of Soli

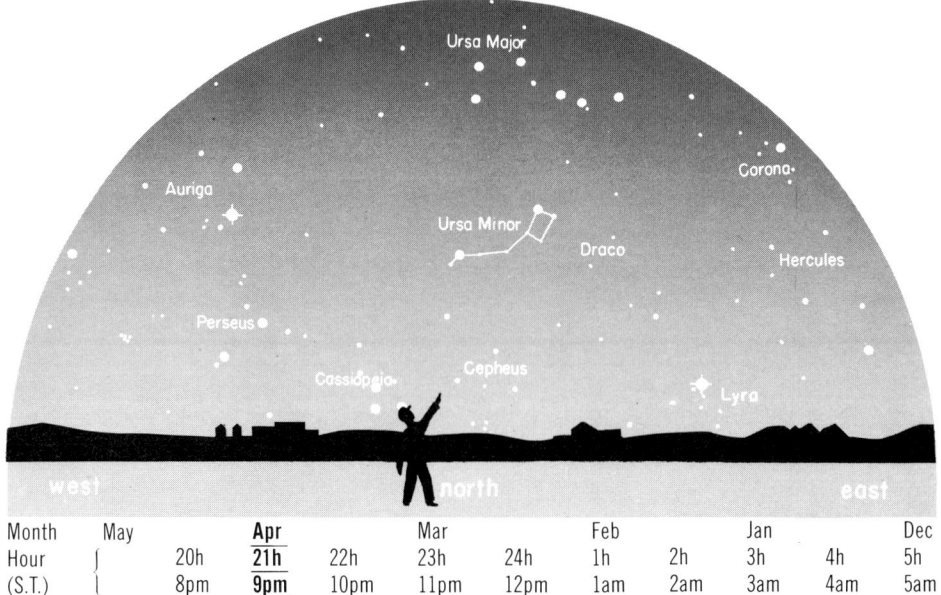

Month	May	**Apr**		Mar		Feb		Jan		Dec
Hour	20h	**21h**	22h	23h	24h	1h	2h	3h	4h	5h
(S.T.)	8pm	**9pm**	10pm	11pm	12pm	1am	2am	3am	4am	5am

1. Face North.
2. Map shows the sky for the months listed and the times given below the months.

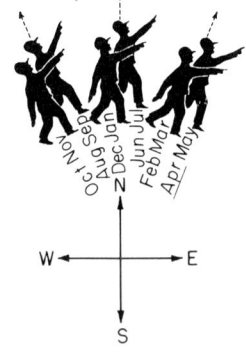

1. Select the month.
2. Face the compass point.
3. Point your arm as high as the Little Man's.

THE LITTLE BEAR

Now that Callisto was Ursa Major and Arcas, Ursa Minor (see page 36), and thus honored by being given a type of immortality, Juno, in her jealousy, became more vindictive. She went to her brother, Neptune, god of the seas, and asked him to minimize the sky honor accorded to Callisto and Arcas, revolving in the northern sky. To appease his sister, Neptune ruled that Mother Bear and Son Bear could never enter his western sea baths, a privilege given to all other star groups. At what latitude must the original story teller have been living in order to make this circumpolar legend sound convincing? Can you see why the body of water at the northern part of the world is called the Arctic Ocean?

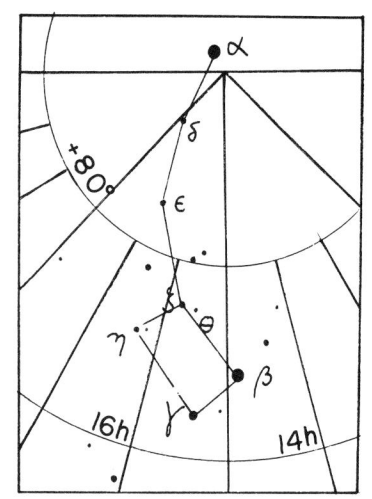

URSA MINOR includes the Pole Star, variously called Polaris or the North Star. It is very close to the north pole of the sky. This point is just over the north pole of the earth and therefore is fixed in its position on the sky. The Pole Star is almost fixed, and to the *naked eye* it is always in the same spot. It points out the north, and by measuring its height above the horizon we can find out our latitude. Polaris is at the end of the handle of the Little Dipper, or the end of the tail of the Little Bear. The two stars at the end of the bowl of the Dipper are the Guardians. They circle around the pole closer than any other bright stars. Trace out the Dipper and notice how it pours into the Big Dipper.

With the Naked Eye. Find the Pole Star, α Ursae Minoris.

With Binoculars. You may be able to catch a glimpse of the eighth magnitude companion to Polaris.

With the Telescope. Polaris is a double star of magnitudes 2 and 8, separation 18″. The second magnitude component is a variable similar to δ Cephei (page 65), changing its brightness periodically by a half magnitude. There is also evidence that there is another close companion revolving around it, closer than the eighth-magnitude star. This system is about 500 light-years away.

PHOTOGRAPHIC TRAILS AROUND THE NORTH POLE

If you mount a camera rigidly and let it face the north pole with the shutter open you will get a series of arcs similar to the ones shown. The longer the exposure the longer the arcs. If the sun did not block out the stars by day with its own bright rays, you could expose the film for 24 hours and get complete circles for each star trail. Should you expose for 12 hours you would then get half-circle trails. Likewise, a 3-hour exposure will give you an arc of $3/24$, or $1/8$, of a full circle. How long do you estimate the exposure in this photograph was? (Photograph by F. J. Chappell, Lick Observatory.)

VIRGO / LIBRA

*"Lo, the Virgin, in her hand a glittering ear of corn . . .
And bore the name of Justice, summoning the elders."*
PHENOMENA, by Aratus of Soli

Month	Jul		Jun		May		Apr		Mar		Feb		Jan
Hour		20h	**21h**		22h	23h	24h	1h	2h	3h	4h	5h	6h
(S.T.)		8pm	**9pm**		10pm	11pm	12pm	1am	2am	2am	4am	5am	6am

1. Face South.
2. Map shows the sky for the months listed and the times given below the months.

1. Select the month.
2. Face the compass point.
3. Point your arm as high as the Little Man's.

THE VIRGIN / THE SCALES

Astraea and her sister, Pudicitia, innocent young goddesses, came down from the heavens to live among the peaceful people of the earth. Wheat, grain, fruit and milk were the diet for all. Not a drop of blood was shed for food or in combat. Astraea was the roving Justice.

Then the pastoral vista changed and Astraea was horrified to see people refusing to abide by her peace plans. They made war; they robbed the traveler; and they oppressed the poor. They became so delinquent that Astraea moved to the mountain glens, but some followed her even into these remote valleys. Astraea's celestial family moved back to the skies, but she bravely remained, believing justice would eventually prevail. When crime continued to increase, she, too, had to leave. Justice disappeared from the earth and there was nothing left but woe and despair. She placed herself in the skies as VIRGO and during the night she shows herself only to those mortals still interested in peace and justice. As an added reminder, Virgo broke a blade of wheat, and, scattering the grains across the sky, formed the peaceful Milky Way.

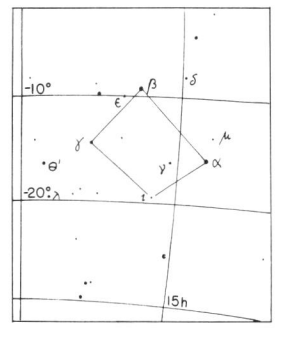

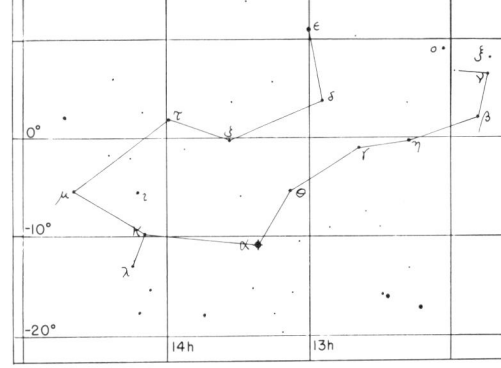

LIBRA, which now means "scales," originally meant "weights." Can you see how Addison, the English essayist, could report on a vivid dream in which he saw the goddess of Justice descending from the constellation to regulate the affairs of men, weighing the evidence in the Balance? This constellation was once a part of Scorpius, forming the claws of the Scorpion.

With the Naked Eye. Spica, α Virginis—a beautiful white star of the first magnitude—is about 150 light-years away. Between β and ε is the Field of the Nebulae where more than 300 of these remote objects were discovered by Herschel. These are now known to be galaxies beyond our Milky Way system. α Librae, "Zuben el genubi," borrowed from Scorpius, represents the southern claw. β Librae, "Zuben es schamali," represents the northern claw.

With Binoculars. Observe Spica, which is more than 1,000 times as bright as the sun. α Librae is a double; magnitudes 3 and 6.

With the Telescope. γ Virginis is a famous double star with a period close to 180 years. Each star is nearly the fourth magnitude, 6″ apart. One of the brightest quasars, known as 3C273, has been found near δ Virginis. Though invisible through a small telescope, this type of object radiates a great deal of energy, the source of which is something of a mystery. Also, quasars are thought by many astronomers to be the most distant objects in the universe. δ Librae, just northwest of β Librae, is a short-period variable star changing from fifth to sixth magnitude in a cycle of about 2.3 days, similar to the case of Algol. (See Perseus, page 74.)

SPIRAL GALAXY SEEN ON EDGE, MESSIER 104

Compare this galaxy with the ones on pages 31 and 71. Galaxies take on many different forms which are partly dependent on the angle at which we view them and partly on the pattern of the galaxy itself. The dark rim is composed of cold dust and gas, which are not transparent. (200-inch photograph, Mt. Wilson and Palomar Observatories.)

BOÖTES

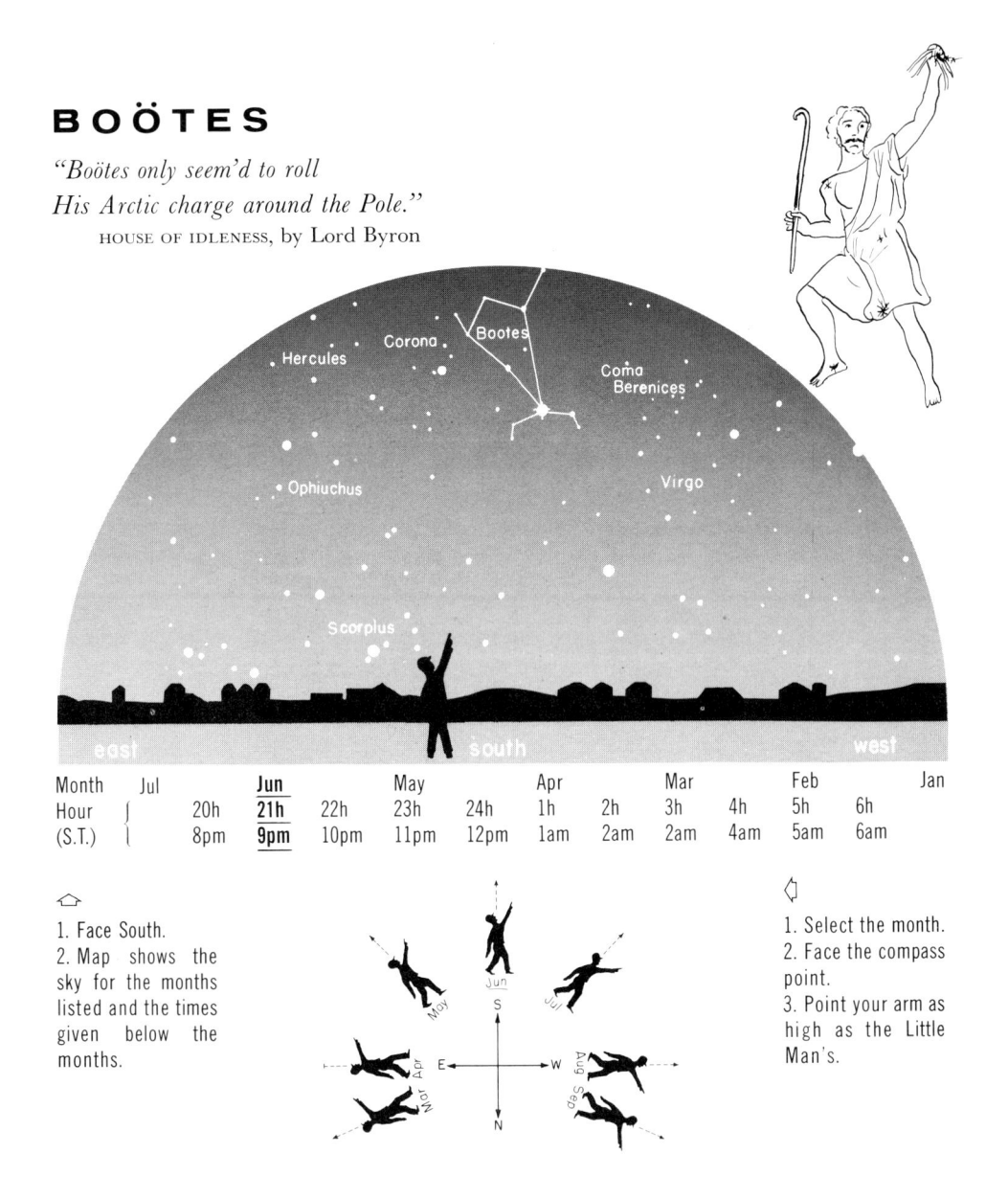

*"Boötes only seem'd to roll
His Arctic charge around the Pole."*
HOUSE OF IDLENESS, by Lord Byron

Month	Jul		Jun		May		Apr		Mar		Feb		Jan
Hour		20h	**21h**	22h	23h	24h	1h	2h	3h	4h	5h	6h	
(S.T.)		8pm	**9pm**	10pm	11pm	12pm	1am	2am	2am	4am	5am	6am	

1. Face South.
2. Map shows the sky for the months listed and the times given below the months.

1. Select the month.
2. Face the compass point.
3. Point your arm as high as the Little Man's.

THE BEAR DRIVER

When Ursa Major and Ursa Minor (see pages 36 and 44) left the earth and went into orbit in the northern skies, they needed someone to guide them and keep them circling the Pole Star. Boötes was chosen for this astronautic role. His task was to make sure that mother and son bears did not leave their orbits and disappear into space. The gods wanted the bear stars up there because their starlight had become a part of a simple navigational system which ship captains depended upon for finding their position and direction at sea.

48

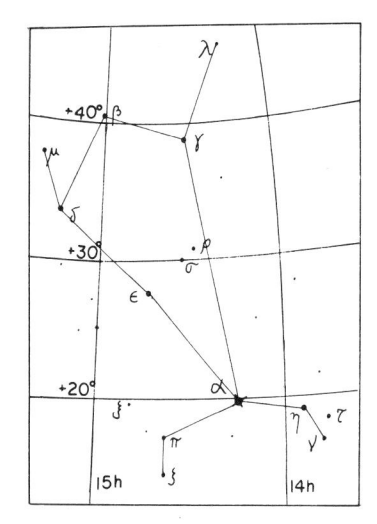

Boötes tied leashes to his two hunting dogs, took his lance, or bear-prod, in hand and began tracking after the two bears. In this manner, he still hounds Callisto and Arcas night and day, keeping them on a circumpolar orbit around the top of the earth, one such trip taking him a little less than 24 hours (see page 12). We wonder why twinkling Arcturus is shown to be a swinging light on the fringe of Boötes' tunic and not on the point of his lance. Could this be the first tail-light man invented?

BOÖTES is a kite-like figure of five stars and is very well known. At the end of the kite's tail is the bright star Arcturus.

With the Naked Eye. Arcturus is α Boötis. This brilliant sun is of deep golden or reddish color and, with Vega and Capella, is one of the brightest stars in the sky. Its distance is about 40 light-years. This "sun" is moving through space at a rate of 75 miles per second with respect to our sun.

With Binoculars. Notice the beautiful groups of dim stars near Arcturus. Get Arcturus just outside the field of view.

With the Telescope. ε Boötis is one of the finest doubles; magnitudes 3 and 5; separation 3″; colors, orange and green. This star is about 180 light-years away. δ Boötis is a double; magnitudes 3 and 7; separation 2′.

ABOVE THE SUN'S SURFACE

The sun's disk is blocked out so that you see the upper atmosphere of the sun. This photograph shows a giant prominence, a region of great surging hot gas rising some 80,000 miles above the sun's surface. Nuclear particles shot out from such prominences, which reach the earth's atmosphere, cause the northern lights seen in high northern latitudes. The white disk on the right is the size of the earth on this scale. (Mt. Wilson Observatory photograph.)

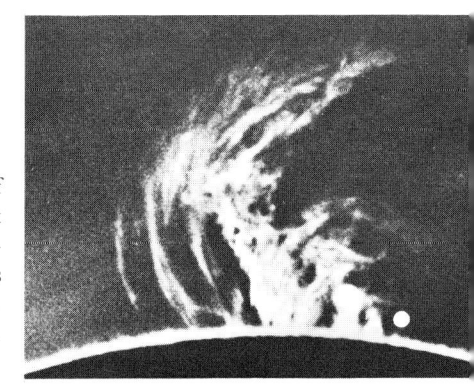

CORONA BOREALIS

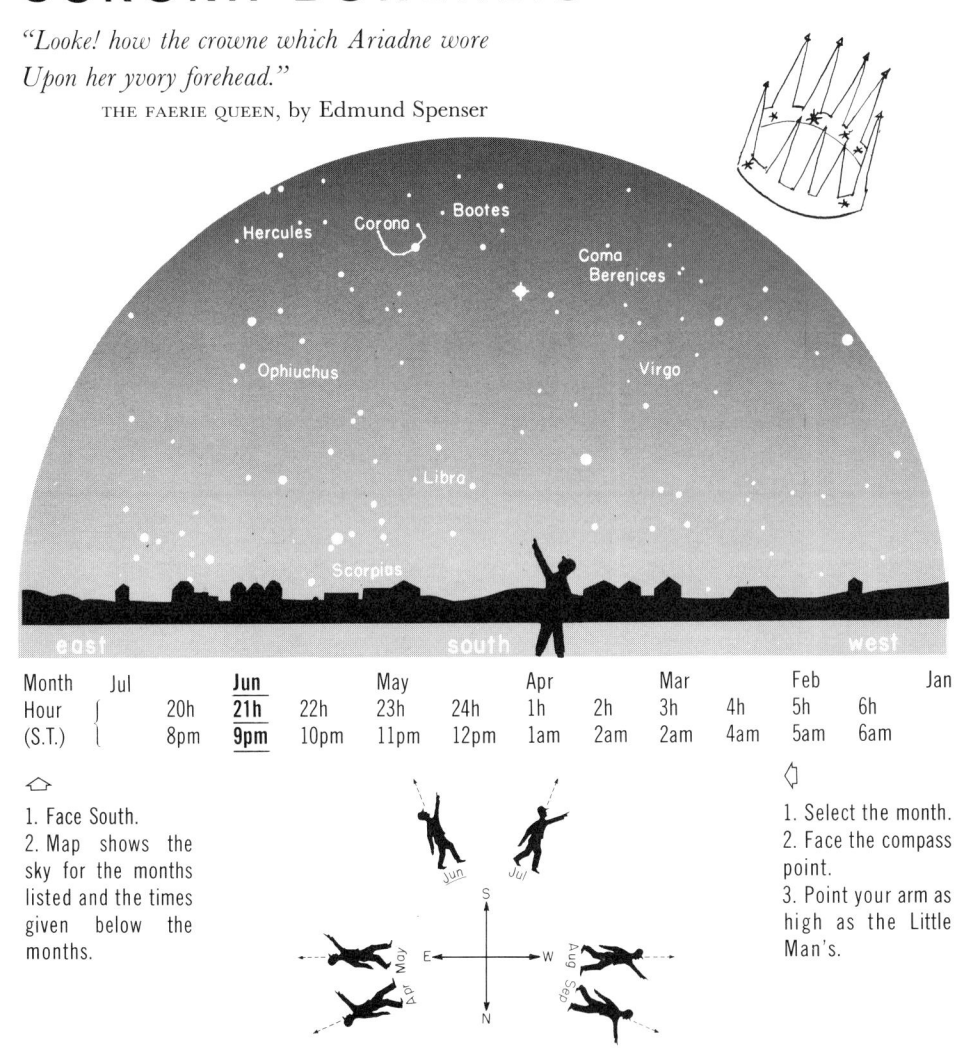

"Looke! how the crowne which Ariadne wore
Upon her yvory forehead."
  THE FAERIE QUEEN, by Edmund Spenser

Month	Jul		Jun		May		Apr		Mar		Feb		Jan
Hour		20h	**21h**	22h	23h	24h	1h	2h	3h	4h	5h	6h	
(S.T.)		8pm	**9pm**	10pm	11pm	12pm	1am	2am	2am	4am	5am	6am	

1. Face South.
2. Map shows the sky for the months listed and the times given below the months.

1. Select the month.
2. Face the compass point.
3. Point your arm as high as the Little Man's.

THE NORTHERN CROWN

Princess Ariadne, beautiful daughter of Minos, King of the island of Crete, fell in love with Theseus, prince of Athens. He had volunteered to be a victim to Minotaur, the monster who was half bull and half man, and a favorite god of the Cretans. Theseus was jailed in the Labyrinth, an underground dungeon. Ariadne, torn between loyal love for her father and country and her powerful love for Theseus, decided to help him escape. She discovered the silken thread which led her through the maze of the Labyrinth to the place where Theseus was being held. As soon as Ariadne had cut Theseus' bonds with a sword, he took the weapon and slew Minotaur. The lovers then fled by sailboat to Noxos, an adjacent island, fleeing the anger of Minos. As the seasonal wind shifted around to the direction of Athens, Theseus sailed away from his sweetheart on Noxos, though planning to return to her. This he never did because many tragic events, including the death of his father, Aegeus, prevented him. Ariadne pined away, waiting for Theseus to sail back for her. Finally, the god Bacchus, coming down from the heavens, comforted Ariadne, made her his bride, placed her in the skies, and gave her a jewelled crown of seven twinkling stars. Do you think that Corona Ariadnes would be a more appropriate name for the lovely half-circle of stars than CORONA BOREALIS?

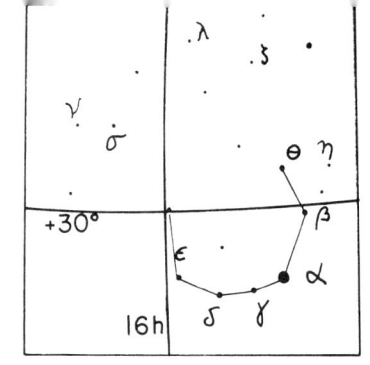

The circle of seven stars is one of the most beautiful of all the smaller groups. One jewel in this crown is brighter than the other six. This star, Gemma, meaning the jewel of the crown, is marked α on the chart; it is between second and third magnitudes. Some see seven Indian chiefs around the council fire in this constellation. Others call it a horseshoe. This is such a striking little figure that one cannot fail to remember and recognize it.

With the Naked Eye. Perhaps you can see that ν Coronae is a double star. It is yellow.

With Binoculars. Study ν Coronae. Note the color. The stars are 10' apart, and both are fifth magnitude, and yellow.

With the Telescope. ζ Coronae is a double star; magnitudes 5 and 6; separation 6"; do they look white and green to you? σ Coronae is a double star; magnitudes 5 and 6; separation 5". A hundred years ago these two stars were much closer together. They revolve around one another in a period of several hundred years.

SOLAR CORONA DURING TOTAL ECLIPSE OF
SUN JUNE 30, 1954

The corona extends high above the visible surface of the sun, beyond the prominences (p. 49). It is seen fully only during total eclipses of the sun, and has been detected out in space to 14 times the solar radius. Small changes in the shape of the corona are continuously taking place. The overall shape differs greatly during the years of many sunspots compared to the years of minimum sunspots. Although you can see the corona on the photograph, its density is far less than that of any vacuum we can produce in a laboratory on the earth. (Photograph: 4-inch refractor by L. Dahlmark.)

LYRA

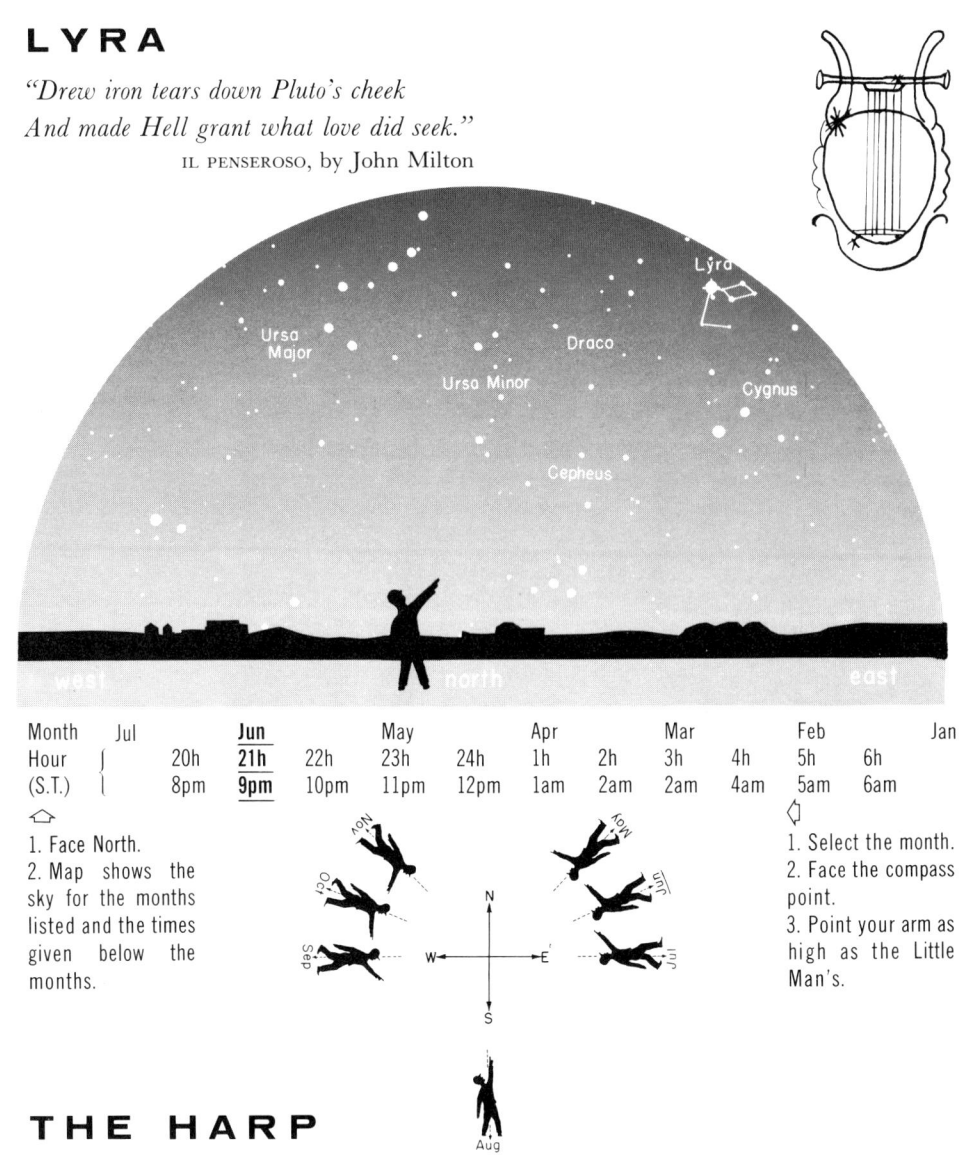

*"Drew iron tears down Pluto's cheek
And made Hell grant what love did seek."*
           IL PENSEROSO, by John Milton

Month	Jul		**Jun**	May		Apr		Mar		Feb		Jan
Hour		20h	**21h**	22h	23h	24h	1h	2h	3h	4h	5h	6h
(S.T.)		8pm	**9pm**	10pm	11pm	12pm	1am	2am	2am	4am	5am	6am

1. Face North.
2. Map shows the sky for the months listed and the times given below the months.

1. Select the month.
2. Face the compass point.
3. Point your arm as high as the Little Man's.

THE HARP

Lyra is the harp which Apollo, god of the sun and of music, lent to his favorite mortal, the musician Orpheus. Apollo thought highly of this harp which his mischievous brother Mercury had invented out of an empty tortoise shell. Mercury had given Lyra to Apollo as a peace offering for having teased him so much. At first Orpheus used Lyra to make exquisite music which charmed all nature. But when Orpheus fell in love with Eurydice, he forgot that Apollo expected him to keep on playing the lyre. Besides, everyone missed the lovely music. Apollo would not tolerate having his precious harp neglected and, to punish Orpheus, he sent a serpent to sting Eurydice to death. Heartbroken Orpheus moaned and mourned, but found no comfort until he returned, playing Apollo's harp. Orpheus followed dead Eurydice into the shadowy nether regions making sad sweet music. Even hard-hearted Pluto, god of the underworld, cried iron tears on hearing the plaintive melodies which Orpheus made by plucking the harp strings. Pluto relented and agreed to permit Eurydice to return to earth under one condition—that when Orpheus started back, he would wait until he was above ground before he checked to see if Eurydice was following him. Up to the halfway mark, Orpheus looked straight ahead, trembling and tempted to turn to see if his sweetheart was there behind him. Beyond this point, however, he, alas, broke his agreement, looked back and saw wispy Eurydice, suspended between a cloud form and a human being, before she disappeared forever. Apollo took back his favorite Lyra and hung it among the stars.

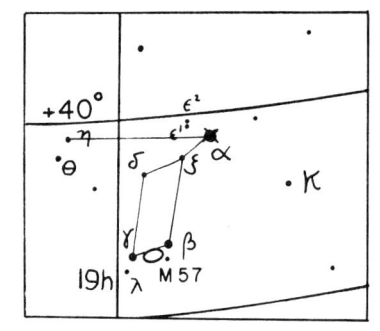

The celestial dome was sometimes known as The Sphere. Can you think of a hymn which tells of the music of the spheres?

LYRA is small but has the bright star Vega, α Lyrae, to make it important. This brilliant white star is second only to Sirius in the north. Close by it are two stars forming a little triangle. The sun appears to be moving through space at 12 miles per second toward a point near this constellation.

With the Naked Eye. Study the triangle of Vega, ε Lyrae, and ζ Lyrae. Observe ε Lyrae carefully and see if you can see it as a double. This is a very rigid test for good eyesight.

With Binoculars. ε Lyrae can easily be seen as a double star. Each one is of magnitude 5, and they are 3½′ apart. Study the variable star β Lyrae. It varies from magnitude 3 to 4 in a period of nearly 13 days.

With the Telescope. ε Lyrae is now four stars. Each pair is of magnitudes 5 and 6 with separations 2″ and 3″. M 57 marks the Ring Nebula. You will see it as a hazy smoke ring. ζ Lyrae is a double; magnitudes 4 and 6; 44″ apart. β Lyrae is a double as well as a variable. It is really four stars forming a Y of 4, 7, 8, and 11 magnitudes; separation 45″, 65″, and 85″.

THE MOON

Photo of lunar surface taken by Surveyor I on June 2, 1966, within a few hours after the spacecraft soft-landed near the moon's equator. A large rock, about six inches high and a foot long, appears at right. A number of small rocks are scattered about a small crater in upper center of photo. Bright spots at left of crater are reflections of the sun in the TV camera system. This 600-scan-line picture was one of 144 TV photos taken by Surveyor I during its first day of operation on the moon. (NASA/JPL)

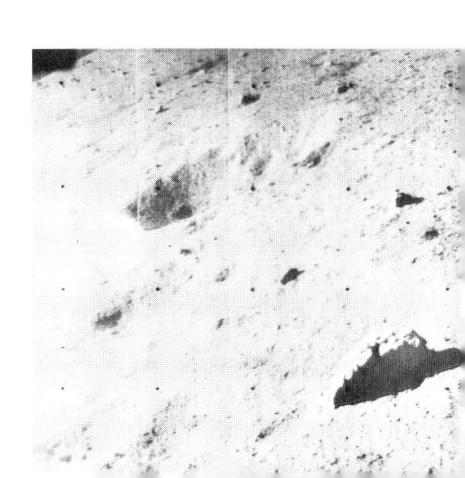

CYGNUS

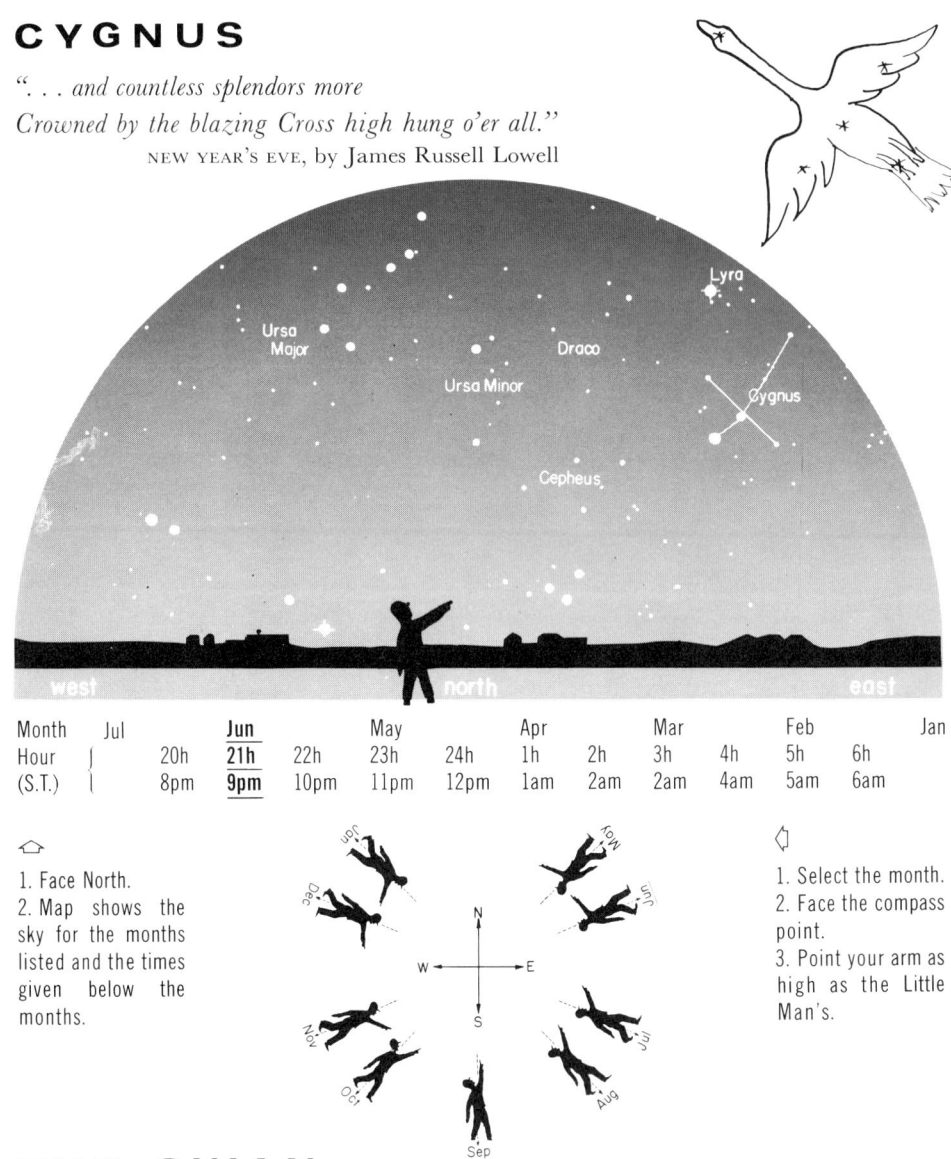

"*. . . and countless splendors more*
Crowned by the blazing Cross high hung o'er all."
NEW YEAR'S EVE, by James Russell Lowell

Month	Jul		Jun		May		Apr		Mar		Feb		Jan
Hour (S.T.)		20h 8pm	**21h** **9pm**	22h 10pm	23h 11pm	24h 12pm	1h 1am	2h 2am	3h 2am	4h 4am	5h 5am	6h 6am	

1. Face North.
2. Map shows the sky for the months listed and the times given below the months.

1. Select the month.
2. Face the compass point.
3. Point your arm as high as the Little Man's.

THE SWAN

The Swan, or Cygnus, was the name given to the mournful friend of Phaeton, the son of Apollo. Young Phaeton persuaded his father to permit him to drive the family sun chariot around the world for just one day. Because the horses had too much power, he lost control, and orbited close to the earth, the chariot smoking and flaming. Watching the approaching accident with horror were Cygnus, his friend, and Jupiter, Father of the Gods. Jupiter side-tracked the young charioteer with a bolt of thunder and Phaeton plunged into the river Eridanus. Cygnus, heartbroken at the loss of his friend, tried to recover

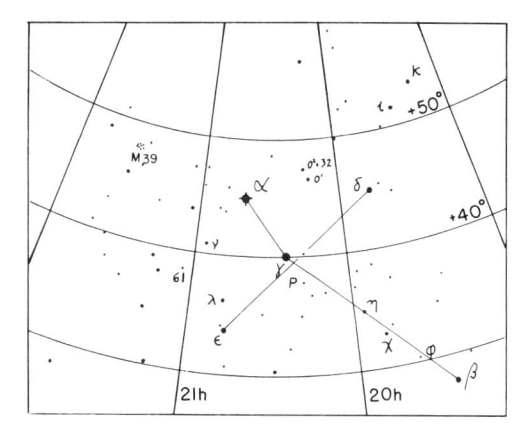

Phaeton's body by diving time after time like a sad swan into the river. The gods, out of pity, turned him into a swan and placed him in the sky. The Swan's journey in the overhead sky occurs about the fall of the year. Isn't this a good accompaniment for the flight of migratory birds before winter?

CYGNUS, in the midst of the Milky Way, contains the Northern Cross. The star at the top of the cross is Deneb, α Cygni. At the foot of the cross is β Cygni, or Albireo. Near Deneb is a dark rift in the Milky Way known as the Northern Coalsack.

With the Naked Eye. Identify α Cygni, Deneb and β Cygni, Albireo and trace out the Cross. Find o Cygni. Can you see a faint star near this one?

With Binoculars. β Cygni is one of the finest doubles in the sky and good binoculars will just separate it. It is better studied with the telescope. Find 61 Cygni. It was the first star to have its distance measured. It is only 11 light-years away, or about 650,000 times as far away as the sun. Study the Coalsack and the rich fields along the Milky Way near Deneb. Examine $o^{1,2}$ Cygni.

With the Telescope. The components of β Cygni are magnitudes 3 and 5; 34″ apart; colors, gold and blue. 61 Cygni is also a double; 5 and 6 magnitudes; 23″ apart; both are golden red. This double star system is one of the closest to us. There is also a third invisible companion which may be something like our planet Jupiter. Examine o^1 and o^2 Cygni. Examine M 39.

NORTH AMERICAN NEBULA

This great gaseous nebula is called the North American Nebula because its shape resembles this continent's appearance on a map. This comparison is most evident in the region corresponding to the Gulf of Mexico. The Gulf of Mexico region is composed of clouds of dust and hydrogen gas which obscure part of the bright cloud beyond. (© and Courtesy National Geographic Society-Palomar Observatory Sky Survey.)

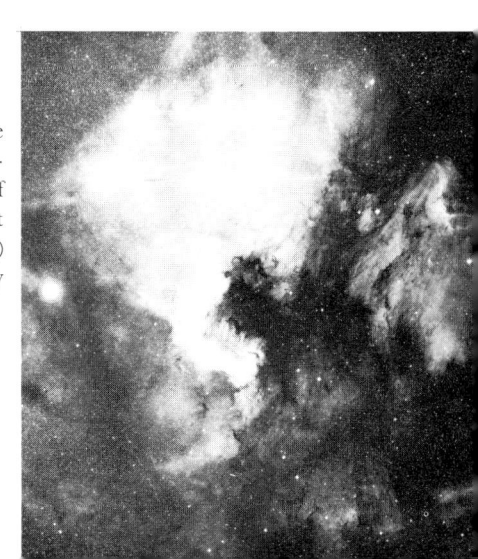

HERCULES

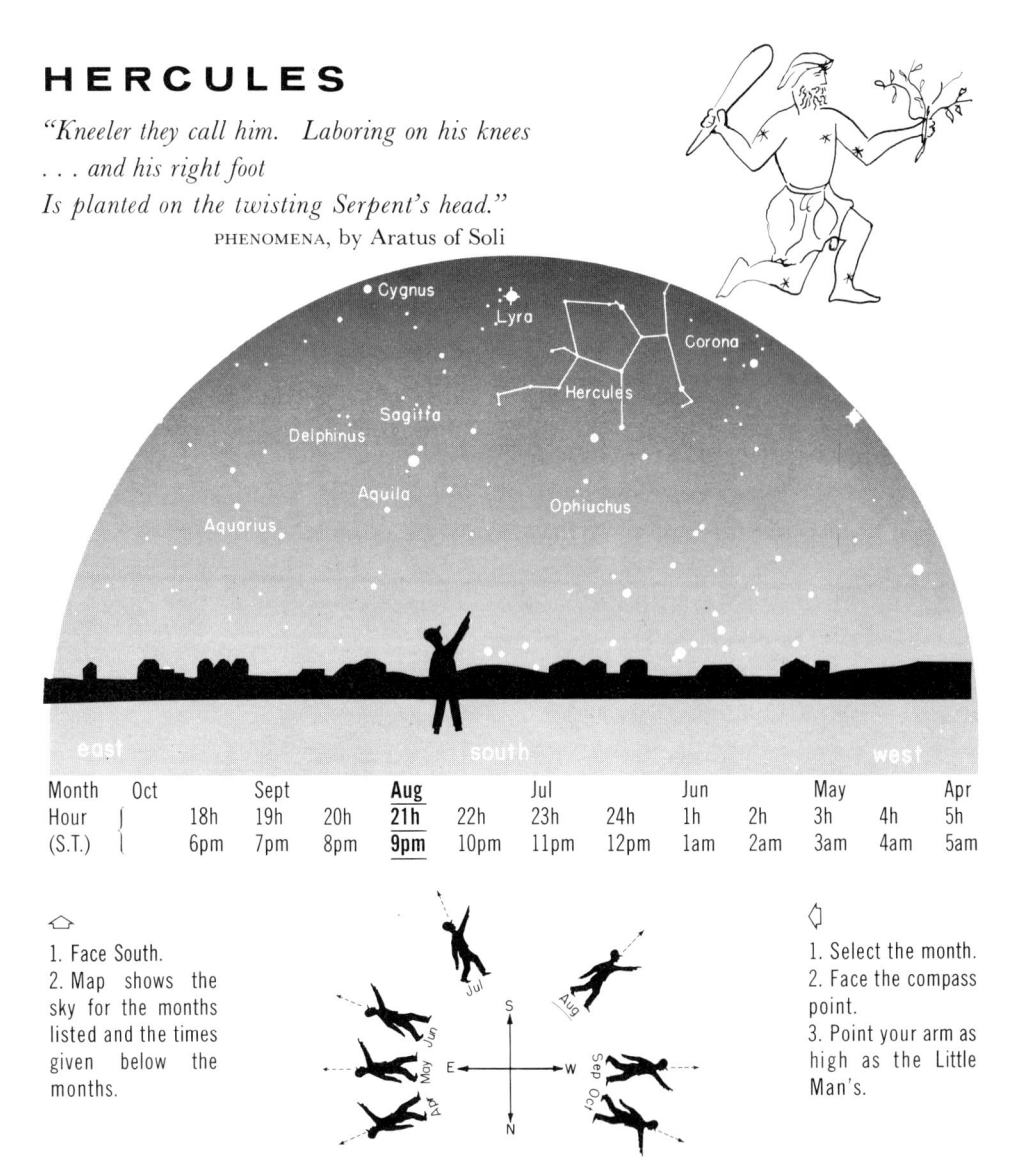

*"Kneeler they call him. Laboring on his knees
... and his right foot
Is planted on the twisting Serpent's head."*
   PHENOMENA, by Aratus of Soli

Month	Oct	Sept			**Aug**	Jul			Jun		May		Apr
Hour		18h	19h	20h	**21h**	22h	23h	24h	1h	2h	3h	4h	5h
(S.T.)		6pm	7pm	8pm	**9pm**	10pm	11pm	12pm	1am	2am	3am	4am	5am

1. Face South.
2. Map shows the sky for the months listed and the times given below the months.

1. Select the month.
2. Face the compass point.
3. Point your arm as high as the Little Man's.

THE STRONG MAN

Several generations after Perseus, the Hero (see page 74), had rescued and married Andromeda, their great-grandson, Hercules, began a life of disasters and exploits of his extraordinary strength. Because Juno, the Queen Goddess, was jealous of the mother of Hercules, she vented her venom on him even when he was a baby by sending two serpents to Hercules' cradle. But his baby hands were so strong that he strangled them before they could poison him. Hercules became a gentle family man, never realizing his own strength. Spiteful Juno overcame this good image of Hercules by making him insane, then commanding him to do away with his three children and his best friend. Although he was judged criminally insane, he still had to pay for his crimes and was sentenced to suffer disgrace and danger. For being gentle, he was ordered to be a slave spinner among the maidens in the court of the Lydian queen for three years—a most embarrassing sentence for a man. For being strong, he was assigned to accomplish twelve labors, among which were the slaying of the Nemean lion (see page 40), and harvesting the apples of the Hesperides, which were guarded by the hundred-headed dragon, Draco. When Hercules heard that Prometheus was unjustly condemned to be chained to the Caucasian Mountains, he climbed these rugged heights and rescued him from the vultures. As old age and impending death approached him, Hercules designed his own funeral pyre. The celestial council, however, sent a heavenly cloud to pick him off the flames and then placed him in the sky. Can you spot the branches of the "old apple tree?" Which foot does Hercules use to stamp on the heads of Draco?

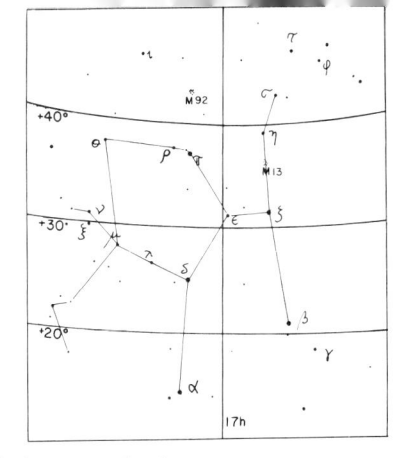

HERCULES is a rather large constellation which is often known as the Kneeler. There are no very bright stars in this group.

With the Naked Eye. If the night is dark and clear, look at the great globular star cluster M 13.

With Binoculars. Study the M 13 cluster and you can see that it is something unusual. The cluster is 30,000 light-years away and contains more than 50,000 stars, about 20 times as many as you can ever see without a telescope.

With the Telescope. Study the M 13 and M 92 globular clusters. M 13 has an apparent diameter of 10'. α Herculis is double; the brighter component is a variable star, while the other component is of the fifth magnitude; colors, red and orange. δ Herculis is double; magnitudes 3 and 8; separation 11". γ Herculis is double; magnitudes 4 and 8; separation 40". ρ Herculis is double; magnitudes 4 and 5; separation 4".

GLOBULAR STAR CLUSTER, MESSIER 13

This concentrated cluster of stars, M 13, is one of about 100 found in our galaxy. There are so many stars toward the center that they appear to blend together. All the stars shown in the photograph are more luminous than our sun. Globular clusters, on the average, contain many more stars and are more nearly spherical than the open, or galactic, clusters. Also, globular clusters contain little or no dust and gas. (200-inch photograph, Mt. Wilson and Palomar Observatories.)

SCORPIUS

". . . that cold animal
Which with its tail doth smite amain the nations."
        Henry Wadsworth Longfellow's translation
        of Dante Alighieri's PURGATORIO

Month	Oct		Sept			**Aug**		Jul			Jun		May		Apr
Hour			18h	19h	20h	**21h**	22h	23h	24h	1h	2h	3h	4h	5h
(S.T.)			6pm	7pm	8pm	**9pm**	10pm	11pm	12pm	1am	2am	3am	4am	5am

1. Face South.
2. Map shows the sky for the months listed and the times given below the months.

1. Select the month.
2. Face the compass point.
3. Point your arm as high as the Little Man's.

THE SCORPION

The ancient council of celestial gods was constantly working to keep a balance of powerful figures so that neither one man nor one beast could become individually omnipotent. When the giant Orion was placed in the skies as an enormous figure, the gods decided to equalize him by creating a mega-scorpion sky figure. As the one constellation makes a prominent appearance in the winter skies, the other is equally honored in the summer skies. Why Scorpius was thus honored is an unanswered question, for he was the monster that frightened the horses of Apollo while young Phaeton (see page 54) was driving them. The sun chariot could have gone completely out of control and burned the earth to a crisp. Originally, Scorpius possessed two gigantic claws and was thus a beastly menace to mankind. Sometime before the first century, these claws were demilitarized, as it were, and made into the constellation of Libra—the scales of justice.

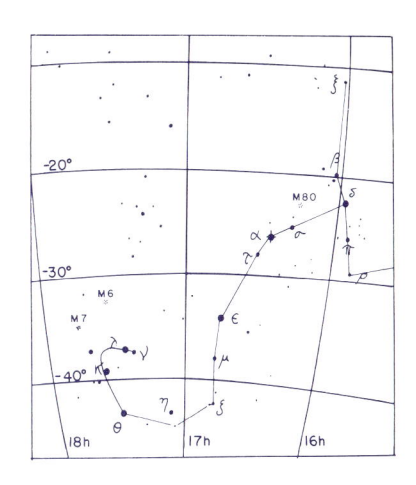

SCORPIUS is a group of the zodiac, yet very little of the sun's path lies within its bounds, less than in any other zodiacal constellation.

With the Naked Eye. Antares, α Scorpii, is a bright red star. The name means Rival of Mars (Anti-Ares). Ares was the Greek name for Mars; and Antares and Mars do look something alike. The diameter of Antares is probably 390,000,000 miles. It is a first-magnitude star and its distance is about 180 light-years. It is more than 700 times as bright as the sun. Notice the pair of stars at the end of the Scorpion's tail. These represent the "sting." In the northern latitudes this constellation never rises high above the horizon so that nearly always the stars appear to twinkle.

With Binoculars. Study the bright stellar region near Antares for many faint stars. Some are double or triple. Look at M 6.

With the Telescope. About midway between α Scorpii and β Scorpii is one of the finest clusters in the sky, M 80. Also look at M 6 and M 7. β Scorpii is a double star, of magnitudes 2 and 4; separation 14″.

JUPITER'S BANDS

If you could look through the 200-inch Hale telescope while holding a piece of red glass in front of the eyepiece, Jupiter would appear as it does in this photograph. A red filter is used to help detect details on the disk of Jupiter. The bands shown on the photograph are made more distinct by this technique. The shadow of the satellite Ganymede is seen on the upper part of the disk. (200-inch photograph, Mt. Wilson and Palomar Observatories.)

SAGITTARIUS

*"With sounding hoofs across the sky I fly
A steed Thessalian with human face."*
    POET'S CALENDAR, by Henry Wadsworth Longfellow

Month	Oct		Sept			**Aug**		Jul			Jun			May		Apr
Hour		18h	19h	20h	**21h**	22h	23h	24h	1h	2h	3h	4h	5h
(S.T.)		6pm	7pm	8pm	**9pm**	10pm	11pm	12pm	1am	2am	3am	4am	5am

1. Face South.
2. Map shows the sky for the months listed and the times given below the months.

1. Select the month.
2. Face the compass point.
3. Point your arm as high as the Little Man's.

THE ARCHER

The ancient Greeks believed Thessalonica, one of their provinces, to have centaurs as inhabitants. These creatures were imagined to have the upper part of a man's body with the rest of the body being that of a horse. Some of these centaurs, the Chirons, were highly educated and were assigned to be the archery teachers to heroes, such as Hercules. Others, Sagittarii, were equipped with bows and arrows and assigned to the defense of the country and to keep the peace. It so happened that Orion, the Hunter, had boasted that he could overcome any beast. The lowly scorpion accepted this statement as a challenge, quietly scurried off to Orion and stung him with a lethal venom. They were both placed in the skies, but widely apart. Then one of the Chiron archers was placed next to Scorpius, the scorpion, to make sure that he would not creep over to the opposite side of the sky and sting Orion again. So we now see the half-horse half-man guarding the peace of the summer night, with his arrow aimed at Scorpius' heart. One move backward into the December skies and the arrow will stop him. Do you think the ancient Greeks were doing a little wishful dreaming to produce a warrior who could run like a horse and shoot arrows like a man?

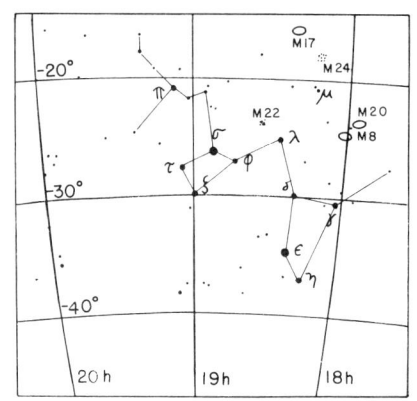

SAGITTARIUS is in the midst of the richest part of the Milky Way, toward the center of our galaxy. Here the great star clouds light the sky in a broad and irregular belt.

With the Naked Eye. Notice the dipper in Sagittarius, called the Little Milk Dipper, probably because it is close to the Milky Way. As seen from the maps it is upside down so the milk is falling out. The cluster marked M 24 should be visible to the naked eye.

With Binoculars. Study carefully the cluster mentioned above and M 8, the Lagoon nebula. John Herschel studied this carefully at his observatory at the Cape of Good Hope. Hunt for, and you will discover, many other clusters and nebulae. Remember that nebulae are great star clouds of dust and smoke, whereas star clusters are predominantly concentrations of stars.

With the Telescope. In addition to studying more carefully the fine clusters noted above, look for the two nebulae M 20 (Trifid) and M 17 (Horseshoe). Also look at the globular cluster M 22. Look at the photograph below and compare it with this region through your telescope.

LOOKING TOWARD CENTER OF MILKY WAY SYSTEM

The telescope is pointed to the richest portion of the Milky Way. In the center is M 8, a very bright nebula which you can see as a faint patch with a small telescope. M 20 is just above. Compare the locations of M8 and M 20 on the photograph with the detailed constellation chart above; also note that the many stars in this region do not seem evenly distributed. Note the "dark lanes" which astronomers believe are composed of dust and gas. Stars on the far side of this interstellar material appear fainter than they would if the space were entirely transparent. (18-inch Schmidt, Mt. Wilson and Palomar Observatories.)

AQUILA / DELPHINUS

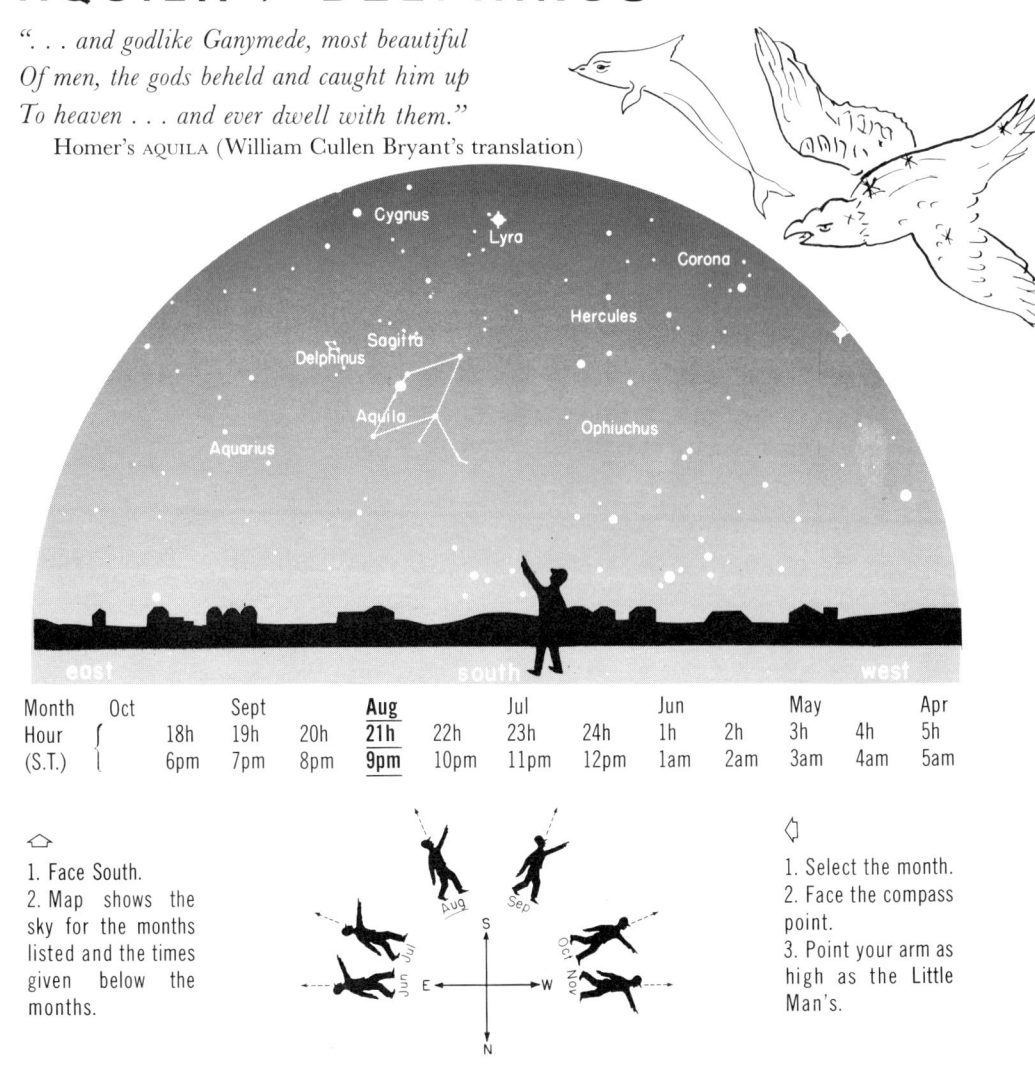

*"... and godlike Ganymede, most beautiful
Of men, the gods beheld and caught him up
To heaven ... and ever dwell with them."*
    Homer's AQUILA (William Cullen Bryant's translation)

Month	Oct		Sept		**Aug**	Jul		Jun		May		Apr
Hour	∫	18h	19h	20h	**21h**	22h	23h	24h	1h	2h	3h	4h	5h
(S.T.)		6pm	7pm	8pm	**9pm**	10pm	11pm	12pm	1am	2am	3am	4am	5am

1. Face South.
2. Map shows the sky for the months listed and the times given below the months.

1. Select the month.
2. Face the compass point.
3. Point your arm as high as the Little Man's.

THE EAGLE / THE DOLPHIN

The Trojans in ancient Asia Minor respected the flying prowess of the eagle, king of birds. They marveled at its powerful flights "into the wild blue yonder," as they saw it disappear high in the sky. No other bird was power-rated as high as Aquila. His fame was known even to the gods and goddesses. Up in the heavens, Jupiter was looking for a replacement for Hebe, the cup-bearer, who had resigned her post because she had tripped several times (and probably spilled nectar over the gods and goddesses!). After surveying all of mankind, Jupiter selected Ganymede, the handsome young son of the Trojan king, as the one who best met the godlike requirements for the position. But how was Ganymede to be transported to the heavenly abode of the gods?

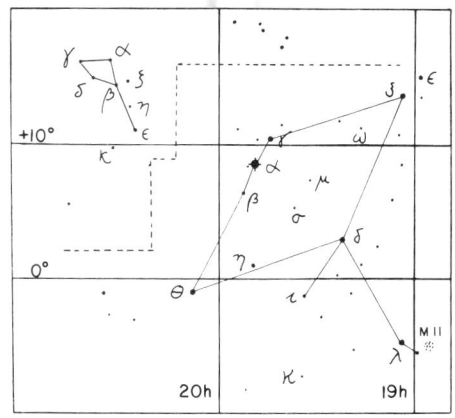

Jupiter remembered Aquila and commissioned him for the project. With Ganymede tucked between his gigantic wings, Aquila took off, rising swiftly into the stratosphere, and delivered his precious cargo right to the throne of Jupiter. As a reward for serving as the vehicle of the first "astronautic" flight, Jupiter immortalized Aquila by placing him in the sky as a constellation in perpetual orbit. You can see him overhead if you live near the middle north latitudes. The American bald eagle perches over the flag of the United States. How many other countries use this symbol of strength and high aspiration?

Neptune, god of the sea, was courting Amphitrite, a goddess of the sea, but she was cool to him. Needing a match-maker, he selected Delphinus, known to sea residents for affability and a soft approach. When Delphinus persuaded Amphitrite to marry the sea-god, Neptune named a small star group for him.

Altair, α Aquilae, and a star on either side of it in a straight line are about all of AQUILA that people generally recognize, but it extends into the Milky Way much farther. DELPHINUS is a diamond-shaped small group and often called Job's Coffin.

With the Naked Eye. Find Altair. This first-magnitude star is one of our real stellar neighbors only 15½ light-years away and shines 10 times as brightly as the sun. Study the variable star η Aquilae. Its average brightness is 4, but it can be a ⅓ magnitude brighter or fainter. The brightness varies with a period of about 7 days. See if you can note this change in brightness. When it is at its faintest, it should have the same brightness as ι Aquilae. Look at γ Delphini and note that it is a close double star.

With Binoculars. Study the variation in η Aquilae. Compare it to other stars in the field, especially ι Aquilae; repeat the observation 3 or 4 days later.

With the Telescope. The cluster near λ Aquilae, named M 11, is very clear.

SATURN

One of the most fascinating sky sights brought to us even with a small telescope is the view of the planet Saturn and its unusual rings. Because of the changing positions of the earth and Saturn in their orbits around the sun, the tilt of the rings changes from year to year. This photograph taken through a large telescope shows that the rings are made up of several parts. Some of these parts are bright, others are faint. Saturn's diameter is nine times that of the earth's, but it is much less dense inside. In fact, Saturn could float in a liquid like our earth water. (100-inch photograph, Mt. Wilson and Palomar Observatories.)

CEPHEUS

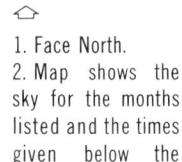

"Kepheus is like one who stretches forth both hands."
                    Robert Brown's translation of ARATUS

Month	Oct	Sept		Aug	Jul		Jun		May	Apr
Hour		18h	19h	20h	**21h**	22h	23h	24h	1h	2h	3h	4h	5h
(S.T.)		6pm	7pm	8pm	**9pm**	10pm	11pm	12pm	1am	2am	3am	4am	5am

1. Face North.
2. Map shows the sky for the months listed and the times given below the months.

1. Select the month.
2. Face the compass point.
3. Point your arm as high as the Little Man's.

THE KING

King Cepheus of Aethiopia had a momentous decision to make. Should he consent to the directive given by the temple oracle of Ammon and sacrifice his lovely young daughter, Andromeda, to the sea monster Cetus? His queen, Cassiopeia (see page 30), had placed him in the position where his role as protector of his people's safety clashed with his feelings as a father. As father of Andromeda, he was aghast at the oracle's edict that only his daughter would be acceptable as a sacrifice to propitiate the gods. As leader of his country, he

64

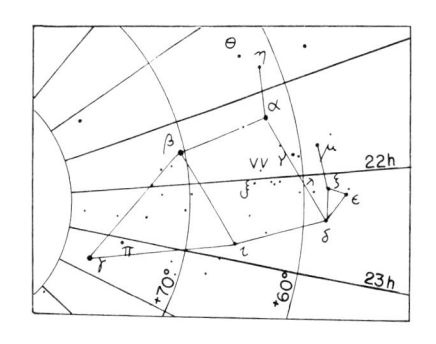

could not make exceptions and substitute another maiden in place of his daughter, else the countryside would be ravaged. Cepheus chose to chain his daughter to the rocks, and left her defenseless. After the crisis over his country had passed (see Perseus, page 74), he was eventually placed in the sky as an inconspicuous constellation. Might Cepheus be considered the symbol of a leader who will appease rather than defend when an ultimatum is delivered?

CEPHEUS is a circumpolar constellation. No star in it is brighter than the fourth magnitude.

With the Naked Eye. Observe the famous variable star δ Cephei for a number of nights and compare its brightness with ζ and ε Cephei, which form a nice triangle. The diagram below explains why we see δ Cephei changing from a faint to a bright star and back again.

With Binoculars. Look at β Cephei, a double; magnitudes 3 and 8, with a separation of 14″. Also notice ξ Cephei, a double with magnitudes 4 and 6, and a separation of 7″.

With the Telescope. Find VV Cephei, a red star of the fifth magnitude. It is known to be a double star whose components eclipse each other every 20 years (see Perseus, page 74). VV Cephei has a diameter about 600 times that of the sun. This fact has been deduced by astronomers from the study of variations in the magnitude of this star.

δ CEPHEI VARIES IN BRIGHTNESS

The envelope of gas around this star expands and contracts in a regular cycle of 5.3 days. Besides the periodic change in size, its brightness also fluctuates with the five-day cycle. If you compare its brightness for a number of nights with that of ζ Cephei and ε Cephei, you can see that δ Cephei sometimes is more nearly the same magnitude as ζ and at other times ε. Many stars pulsate like δ Cephei and are called Cepheid variables. The graph shows the varying magnitude of δ and the constant (straight line) light of ζ and ε.

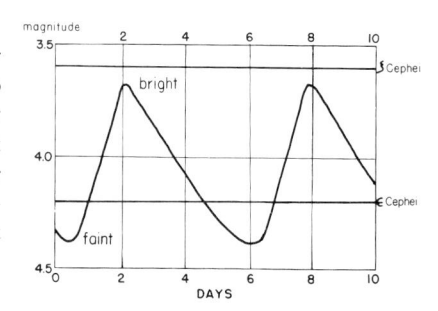

ANDROMEDA

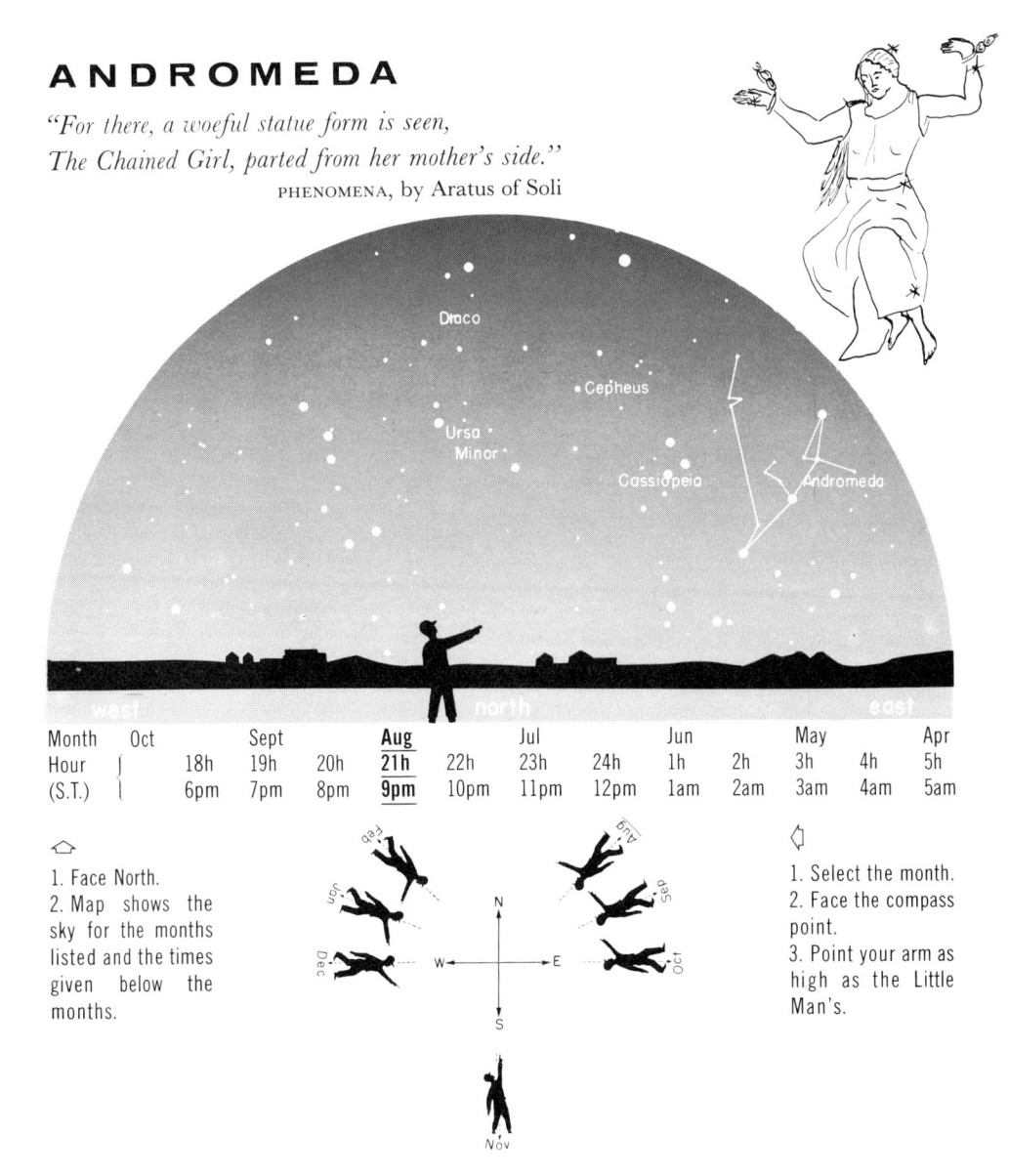

"For there, a woeful statue form is seen,
The Chained Girl, parted from her mother's side."
           PHENOMENA, by Aratus of Soli

Month	Oct	Sept	**Aug**	Jul	Jun	May	Apr
Hour (S.T.)	18h / 6pm	19h / 7pm	20h / 8pm	**21h / 9pm**	22h / 10pm	23h / 11pm	24h / 12pm	1h / 1am	2h / 2am	3h / 3am	4h / 4am	5h / 5am

1. Face North.
2. Map shows the sky for the months listed and the times given below the months.

1. Select the month.
2. Face the compass point.
3. Point your arm as high as the Little Man's.

THE WOMAN CHAINED

The Nereids were the bathing beauties of the ancient world along the Mediterranean Sea. As sea nymphs and goddesses, they felt superior to any mortal beauty. When news reports reached them that Queen Cassiopeia (see page 30) dared to compare the loveliness of her daughter, Princess Andromeda, to their ethereal charms, the Nereids became envious and angry. Appealing to the celestial authorities, they obtained a ruling which sent a huge sea monster to ravage the coast of the kingdom of Cepheus, father of Andromeda. The only salvation for the kingdom—so ruled the temple oracle—was to sacrifice

66

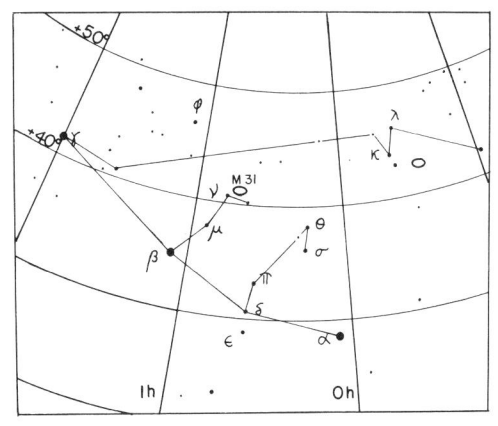

Andromeda to the monster. So the heroine, sobbing and hysterical, was chained to the sea rocks. She watched with horror as the ferocious head of the monster rose high above the waves and came closer to her. The rocks behind Andromeda must have acted as a gigantic sound reflector of her piteous cries for help, for Perseus, the Hero, was able to pick up her SOS from far off. Zooming to the rescue, Perseus (see page 74) released Andromeda and returned her to her parents. Of course the grateful heroine married the handsome, rescuing hero.

At the present time ANDROMEDA might be considered headless, as the star representing her head was renamed Alpheratz—the horse—and is used to complete the Square of Pegasus. Could it be that the Nereids' vengeance was not to be escaped?

The most important object in this group is the Andromeda Nebula, a great galaxy of stars probably 2 million light-years away. This one, nevertheless, is our nearest spiral galaxy.

With the Naked Eye. If the sky is dark, you probably can just see the galaxy, M 31, as a cloudy spot. This is the most distant object you can see with your naked eye. To find it, start from α Andromedae and proceed to the third of the stars in the line of bright stars, then in a direction perpendicular to this line, to the second of the two fainter stars, and look for the galaxy near it.

With Binoculars. α Andromedae, Alpheratz, is second magnitude and about 250 light-years away. Study the Great Nebula, M 31.

With the Telescope. γ Andromedae is double; magnitudes 2 and 5; spaced 10″ apart. π Andromedae is also a double; magnitudes 4 and 8; separation 36″. Be sure to look at M 31.

A GALAXY SEEN ON EDGE,

NEW GENERAL CATALOGUE NO. 891

This is a spiral galaxy which is probably not unlike the one pictured in Ursa Major, but we see it from a different viewpoint. The many stars distributed all over the photograph are in our own galaxy, through which we must look in order to see this and all other galaxies. (60-inch photograph, Mt. Wilson and Palomar Observatories.)

CAPRICORNUS / AQUARIUS

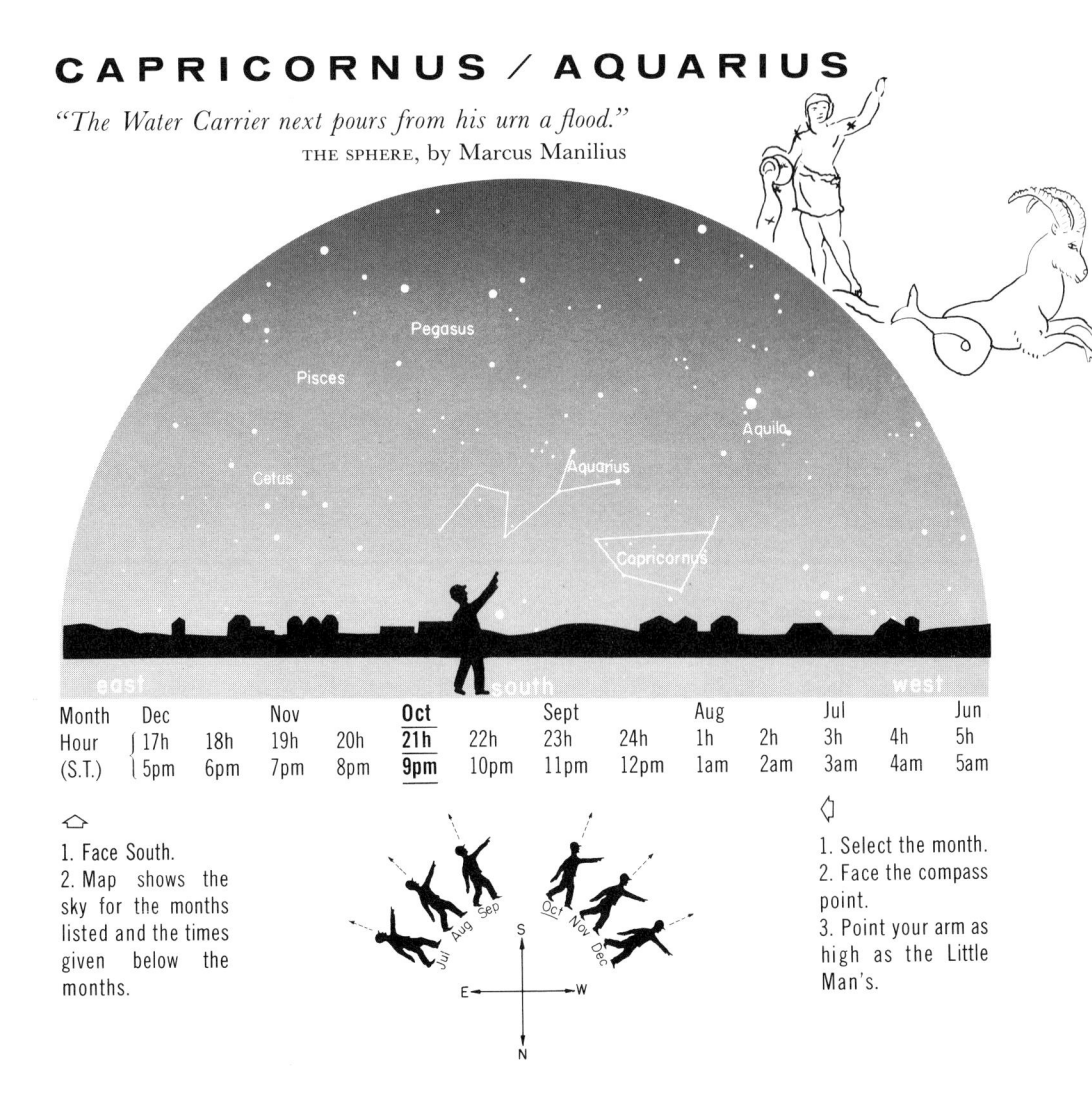

"The Water Carrier next pours from his urn a flood."
THE SPHERE, by Marcus Manilius

Month	Dec		Nov		**Oct**		Sept		Aug		Jul		Jun
Hour (S.T.)	17h / 5pm	18h / 6pm	19h / 7pm	20h / 8pm	**21h / 9pm**	22h / 10pm	23h / 11pm	24h / 12pm	1h / 1am	2h / 2am	3h / 3am	4h / 4am	5h / 5am

1. Face South.
2. Map shows the sky for the months listed and the times given below the months.

1. Select the month.
2. Face the compass point.
3. Point your arm as high as the Little Man's.

THE SEA GOAT / THE WATERMAN

Ganymede, a cupbearer to the gods (see page 62), was empowered by Jupiter to be water-boy to mankind. Tilting his water urn downward, Ganymede poured water endlessly and forever upon the earth, causing rivers to be formed. One of the rivers formed in this way was the Egyptian Nile. The water of the river rose and flooded the dry countryside like an automatic irrigation system. At seasonal intervals, another water boy, Aquarius, coming down from the celestial abode, refilled his huge empty urn by dipping it in the Nile waters. The Nile was the haven for the god, Pan, the goat-footed flute player. On one occasion, the monster Typhon approached Pan and caused him to panic. Pan had no place to find protection so he dived into the Nile. Since his goat hooves were of little use to him in the water, the gods saved him by changing the lower half of his body into a fishtail. His new name was CAPRICORNUS, the sea goat.

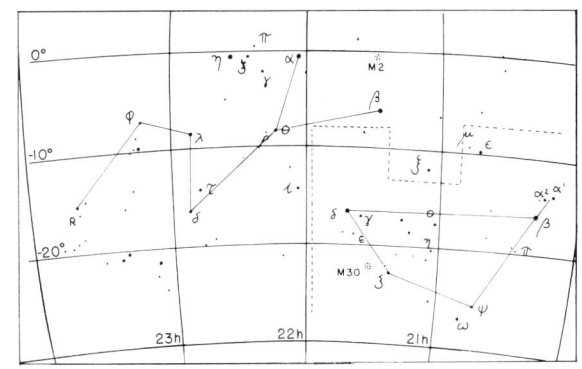

AQUARIUS forms the letter Y. This marks the water jug from which the water carrier is pouring water into the mouth of the southern fish—a strange pastime! Many years ago the sun was in Capricornus about December 23, at the winter solstice. This is the time when the sun is farthest south of the equator, but the point has moved now into Sagittarius. After this date the sun begins to climb the sky again, like a mountain goat. The line on the earth where the sun is just overhead at noon on this date is still called the Tropic of Capricorn.

With the Naked Eye. Find the Y, the water jug; also note the line of stars running south, representing a water stream. You can see that α Capricorni and β Capricorni are doubles. The stars forming α are about 6' apart and those in β Capricorni nearly 4' apart. Just below β Capricorni is a beautiful little group of dim stars.

With Binoculars. Study the stream of stars representing the flowing water. You will find many interesting star groups in it. Also study the two stars at α Capricorni. They are not parts of a true double star, but are drifting apart rapidly. The ancients make no mention of the double because they were too close to be seen, except as one star. β Capricorni is a beautiful double—one star being blue and of magnitude 6 and the other yellow, of magnitude 2.

With the Telescope. ζ Aquarii in the middle of the Y is a double star, both magnitude 4, almost 3" apart. ψ Aquarii is a double of magnitudes 4 and 8, separation 50". α Capricorni—each star noted above is a double. Globular cluster M 2 is about the sixth magnitude and M 30 is about eighth magnitude and is therefore more difficult to see.

PLANETARY NEBULA IN AQUARIUS,
NEW GENERAL CATALOGUE NO. 7293

A great expanding ring of hot gas surrounds a faint central star. In this case the central star is so hot that it sends out powerful X-rays, which make the ring glow something like a fluorescent lamp. It is about 600 light-years away. (200-inch photograph, Mt. Wilson and Palomar Observatories.)

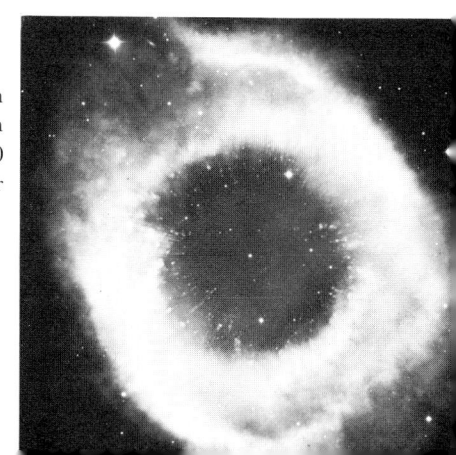

PEGASUS

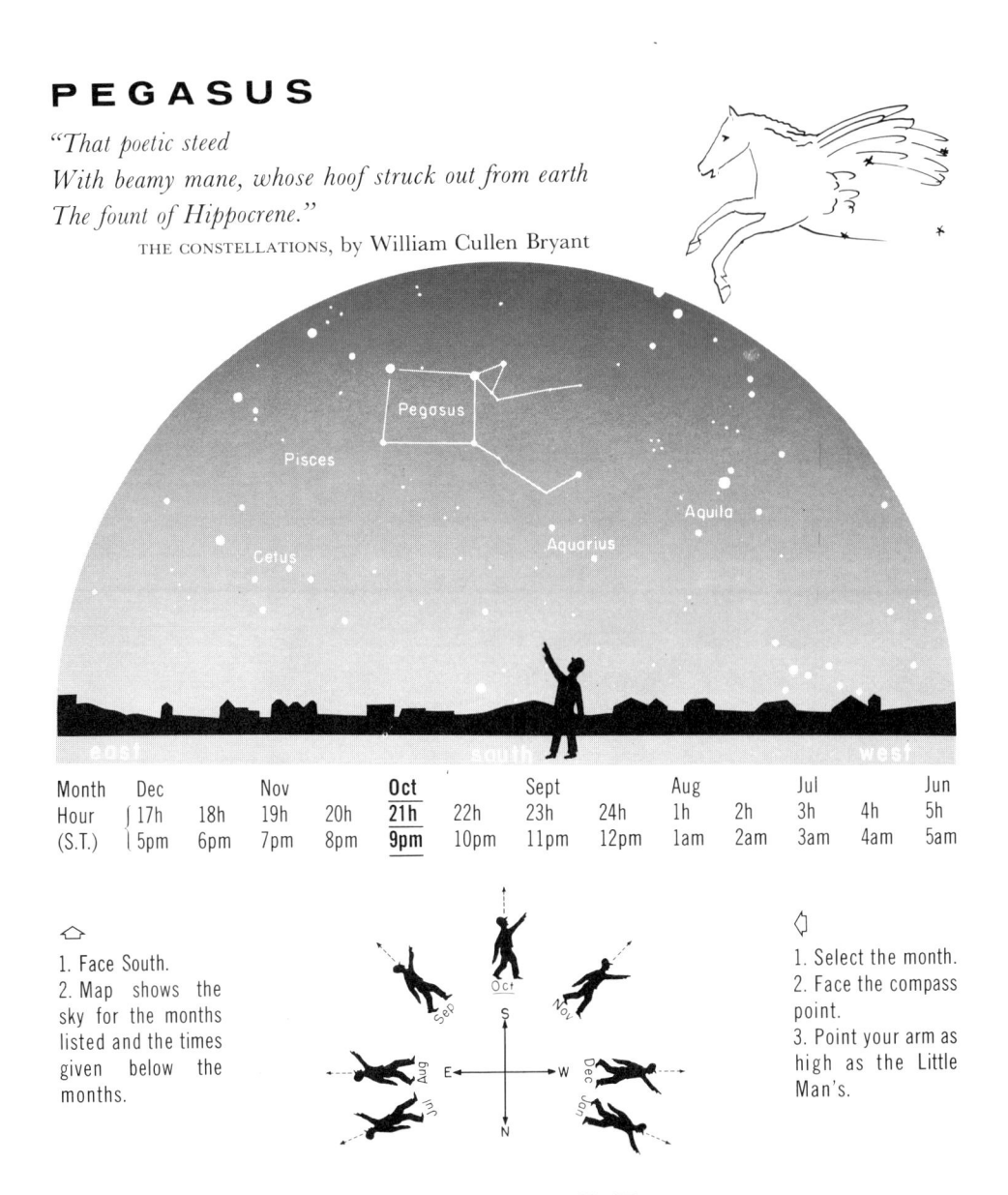

*"That poetic steed
With beamy mane, whose hoof struck out from earth
The fount of Hippocrene."*
        THE CONSTELLATIONS, by William Cullen Bryant

Month	Dec		Nov		**Oct**		Sept		Aug		Jul		Jun
Hour	17h	18h	19h	20h	**21h**	22h	23h	24h	1h	2h	3h	4h	5h
(S.T.)	5pm	6pm	7pm	8pm	**9pm**	10pm	11pm	12pm	1am	2am	3am	4am	5am

1. Face South.
2. Map shows the sky for the months listed and the times given below the months.

1. Select the month.
2. Face the compass point.
3. Point your arm as high as the Little Man's.

THE WINGED HORSE

Neptune, god of the sea, was lolling on his ocean waves when he saw Perseus (see page 74) flying back from a dangerous mission. Perseus was carefully shielding the snake-head of Medusa from his own vision. It was a trophy he prized highly, but he could not stop her blood drops from falling into the ocean below. Neptune was saddened to see that Medusa was dead, as he had loved her in their younger days and wanted to remember her as being beautiful, not evil, as she had turned out to be later. He decided to intervene in the situation and refused to permit Medusa to end in ignominy. Catching drops of her blood and scooping up a handful of sea foam, he carefully molded a snow-white horse, added a pair of wings and launched the steed from the top of the waves. Since this steed, named Pegasus, was thus descended from Medusa and handmade by Neptune, he took off for Olympus, the abode of the gods. Who has not noticed that Pegasus, like the magic carpet of the Arabian Nights stories, is used in advertising as one of the symbols of modern transportation?

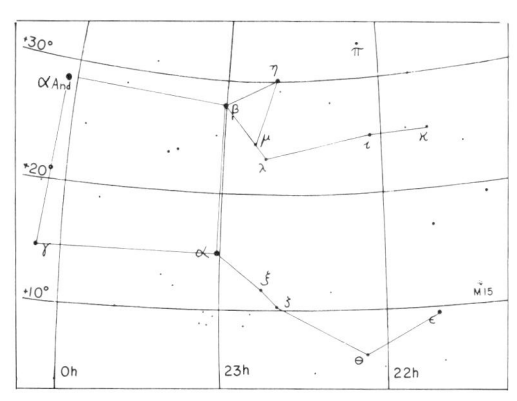

This group is better known for the stars that are not there. The Great Square in PEGASUS is a large blank space nearly 15 degrees square. There are some very faint stars in it, but you will have to look for them. A person with very good eyesight has counted as many as 30, but anyone can find 15 or 20 if the night is dark. One star (at the northeast corner) is borrowed from ANDROMEDA to make the Square. You notice the lines do not quite connect with it in the detailed map.

With the Naked Eye. Count the number of stars you can see in the Great Square. π Pegasi is a double star that you may be able to see as two. Identify Markab, α Pegasi. It is nearly 100 light-years away and is 67 times as bright as the sun.

With Binoculars. π Pegasi can be easily separated into two stars. Sweep over the Great Square to see how many more dim stars you can find. β Pegasi is red and varies slightly in magnitude.

With the Telescope. ε Pegasi is a double star; magnitudes 2 and 8; 2'.5 apart. Notice cluster M 15, discovered by Miraldi in 1745.

BARRED SPIRAL GALAXY

You can see some spiral structure in this galaxy, New General Catalogue 7741, but you also notice the bar across the center. Galaxies show a variety of patterns. There are essentially three main types: irregular, spiral and elliptical. (200-inch photograph, Mt. Wilson and Palomar Observatories.)

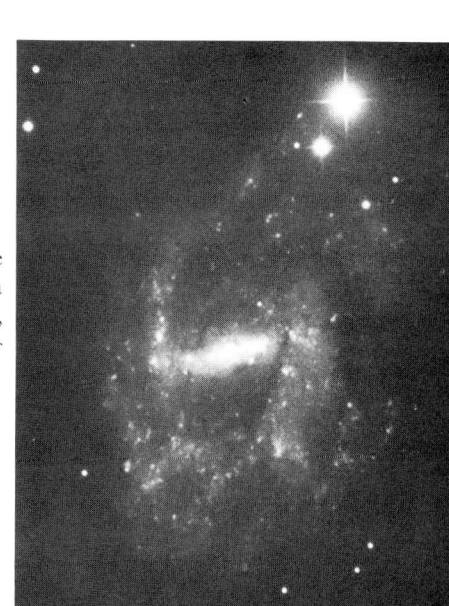

PISCES / ARIES

*"First golden Aries shines who whilst he swam
Lost part of's Freight, and gave the sea a name."*
  THE SPHERE, by Marcus Manilius

Month	Dec		Nov		**Oct**		Sept		Aug		Jul		Jun
Hour	17h	18h	19h	20h	**21h**	22h	23h	24h	1h	2h	3h	4h	5h
(S.T.)	5pm	6pm	7pm	8pm	**9pm**	10pm	11pm	12pm	1am	2am	3am	4am	5am

1. Face South.
2. Map shows the sky for the months listed and the times given below the months.

1. Select the month.
2. Face the compass point.
3. Point your arm as high as the Little Man's.

THE FISHES / THE RAM

Aphrodite, Grecian goddess of love, and her son Eros were vacationing in the vicinity of the Euphrates River. Unexpectedly, the monster Typhon attacked and frightened them. To save themselves, mother and son went down into the river and asked the fishes for protection. Two fishes, Pisces, permitted Aphrodite and Eros to hide behind their scaly bodies. To make sure the frightened pair were not separated in this underwater shelter, the gods joined the fishtails with a flaxen cord.

Helle and her brother Phrixus were the children of Athamas, who gave them a cruel stepmother in the person of Ino. The brother and sister cherished a pet ram given to them by their own mother. When they decided to leave Ino's house, Phrixus and his sister took the ram with them. Wishing to put as much distance as possible between themselves and cruel Ino, they finally reached the sea. No vessels were available for them, however, so they climbed on the ram and started across the waves. Helle became tired, slipped off the ram and was drowned. Phrixus managed to hang on until he reached the shore. Saddened at the loss of his sister, yet grateful that he was saved, Phrixus went to the temple in the grove of Ares. There he sacrificed the ram and hung its hide in the grove. The fleece turned to gold and became worthy of a treasure hunt by the Argonaut, Jason. The sea waters where Helle was lost are named Hellespont.

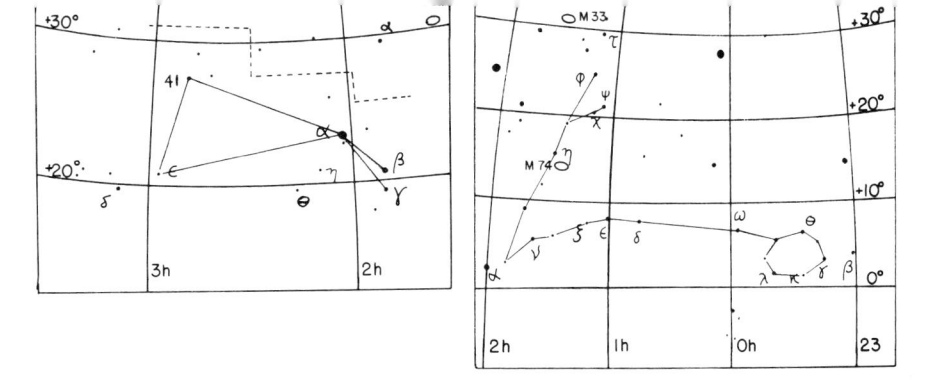

Can you see how this faint constellation is remembered? Hebrew history records that the sun was in ARIES when the Israelites left their Egyptian bondage. Many Christian churches sing of Agnus Dei, the Lamb of God.

It was in the constellation of PISCES that Mars, Jupiter and Saturn were very close together in the year 6 B.C., and some connect this striking combination of planets with the Star of Bethlehem.

With the Naked Eye. The circlet of stars that is small and under the Square of Pegasus is the Western Fish. At the junction of the line joining the two fishes is the brightest star in the group, Al Rischa, α Piscium, a name meaning the Knot. Notice the triangle of stars in the Ram's head. The brightest star, α Arietis, is named Hamal; β Arietis, or Sheratan, is about 35 light-years away.

With Binoculars. In Pisces, follow the circlets and the lines of stars connecting them and note the many beautiful little groups of faint stars that you cannot see with the naked eye.

With the Telescope. γ Arietis is a double star, each of which is of magnitude 5; separation 8″. Al Rischa is a fine double; magnitudes 4 and 5; separation 3″; colors, green and blue. ζ Piscium is a double with both magnitudes 6; separation 24″. Note M 33 which appears at the seventh magnitude. Hunt for a nebula at the place marked M 74.

SPIRAL GALAXY IN PISCES, MESSIER 74

M 74 is a beautiful example of a multiple-arm spiral galaxy. We find the same distribution of star colors in our own galaxy. The brightest stars toward the center of our galaxy are red, while the most luminous ones in the region of the spiral arms are blue-white. (200-inch photograph, Mt. Wilson and Palomar Observatories.)

PERSEUS

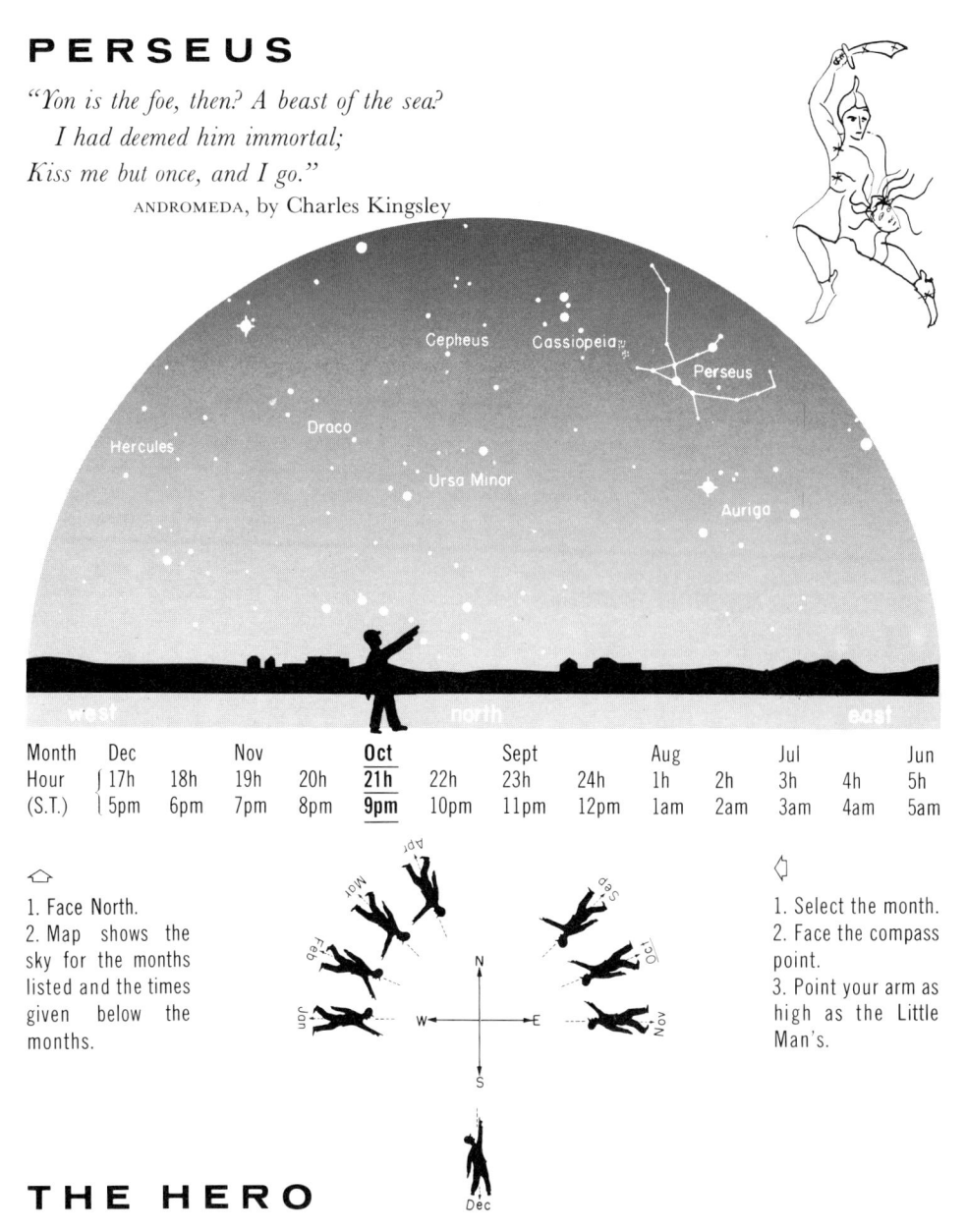

"Yon is the foe, then? A beast of the sea?
  I had deemed him immortal;
Kiss me but once, and I go."
        ANDROMEDA, by Charles Kingsley

Month	Dec		Nov		**Oct**		Sept		Aug		Jul		Jun
Hour	17h	18h	19h	20h	**21h**	22h	23h	24h	1h	2h	3h	4h	5h
(S.T.)	5pm	6pm	7pm	8pm	**9pm**	10pm	11pm	12pm	1am	2am	3am	4am	5am

1. Face North.
2. Map shows the sky for the months listed and the times given below the months.

1. Select the month.
2. Face the compass point.
3. Point your arm as high as the Little Man's.

THE HERO

As a handsome youth, Perseus had the tremendous task of rescuing the lovely maiden Andromeda, chained to the rocks as a sacrifice to the sea monster, Cetus. Even though he needed a weapon more powerful than his sword for this task, undaunted, he travelled to the lair of Medusa. Medusa's head was covered with snakes instead of hair, and the sight of her petrified anyone who looked at her. To guard Perseus against this danger, Minerva had given him her shield, which he used as a mirror so that he would not have to gaze at

74

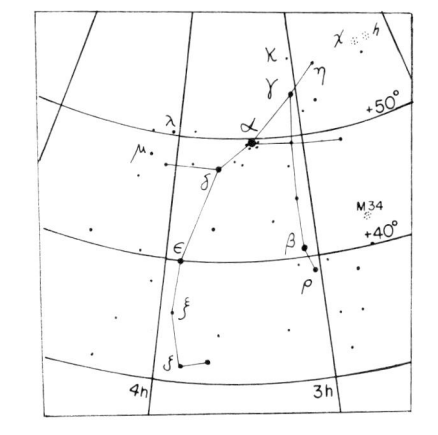

Medusa herself. Approaching Medusa in this way, he was able to cut off her head. He returned to the chained Andromeda and freed her by using Medusa's head to petrify Cetus, the sea monster, who was just about to devour her. Can you see how the ghouls of the Arabian Nights stories might be associated with the star representing Medusa's head?

As the outline guide at the top of this page shows, the lines connecting the various stars of this constellation could be perceived as the letter A. PERSEUS contains the most interesting variable star, the celebrated Algol, or Demon Star.

With the Naked Eye. Keep watching β Persei, Algol, to see if you can see it at its minimum brightness. Find the double cluster. You can just see it on a dark night. It is marked χ and h.

With Binoculars. Observe the double cluster. Look at the cluster M 34, which is the sixth magnitude.

With the Telescope. Study the double cluster. It contains hundreds of stars. η Persei is a double; magnitudes 4 and 8; 28" apart. Study the other cluster marked M 34. It contains over 100 stars.

ALGOL—AN ECLIPSING DOUBLE STAR

Just as we see that the moon eclipses the sun, we can also observe one star of a double star system eclipsing the other. Algol is a good example. When the large relatively cool star eclipses the bright hot star, the total brightness of Algol grows less. The eclipses of Algol do not last long because the star which passes in front of the other moves rapidly in its orbit.

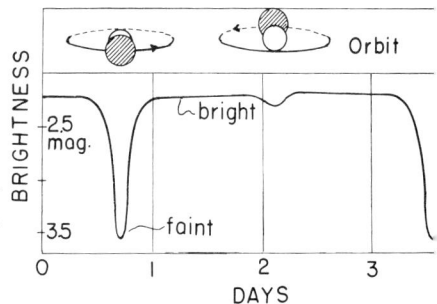

The stars which pass through your zenith

We conclude this chapter with a map, the World Star Chart, which shows all the constellations at one time, something you can never see in the sky, at least not while you are on the earth. The chart is a flat projection of the celestial sphere, which distorts the constellations near the poles in the same way that a flat map of the entire earth shows Greenland, for example, out of proportion to the countries south of it.

You will recall that latitude, to a geographer, means the location of a place on the earth north or south of the equator, and that longitude means the location of a place east or west of Greenwich. A similar system of locating sky objects is used in astronomy, except that these circles are named *right ascension* (R.A.) and *declination* (Dec.). Up to this time, your location of sky objects has been limited to altitude and azimuth angles.

In geography, the zero point for longitude is Greenwich; in astronomy, the zero point for R.A. is the vernal equinox in the constellation of Pisces, the Fishes (see page 72). Going from this point west to east among the constellations, R.A. is measured in hours, minutes, and seconds from 0^h to 24^h. On the World Star Chart the hours will be found at the top and bottom of the page. There are also five faint lines between intervals. It is, of course, impossible to note the second differences on this type of chart.

Again, in geography, the zero point for determining north or south latitude is the terrestrial equator; in astronomy, this is the celestial equator, or equinoctial, found in the center of the chart. Instead of latitude, the name declination (Dec.) is used, with points on this equator being 0 degrees (0°). North declination includes all objects north of the equinoctial and is labeled (+), for example, +20° Dec. South declination means all objects south of the equinoctial and is labeled minus (−), for example, −20° Dec. Declination runs from +90° to 0° to −90°. These numbers are found on the right and left edges of the chart. There are four faint lines between each 10-degree Dec. which are aids to approximating 2-degree intervals in the sky.

On the World Star Chart you can find the individual stars or constellations which pass overhead through your zenith during the year. To do this, find the horizontal line whose declination has the same numerical value as your latitude. The stars along this line are the ones which pass over your head. For example, if you live in New Orleans, Louisiana, or in any community at 30° North Latitude, the bright stars

Castor and Pollux (declination about 30° N.) will pass directly overhead. If you wish to see Vega in Lyra (declination about 40° N.) passing over your head, you will have to move to such cities as Philadelphia, Pennsylvania, or Denver, Colorado, which are at 40° North Latitude. The star at the end of the handle of the Big Dipper, Alkaid (declination about 50° N.), will pass overhead at Winnipeg, Manitoba, Canada, and at all other places at 50° North Latitude.

Suppose you were in a lifeboat in the North Atlantic Ocean and wanted to know your latitude. Looking overhead, you are able to sight Vega passing through your zenith. By remembering that Vega is about +40° declination you can be sure that your latitude is about 40° North. You can see that being able to find the zenith stars is fundamental to the principles of navigation.

No. A-77

WORLD STAR

NORTH

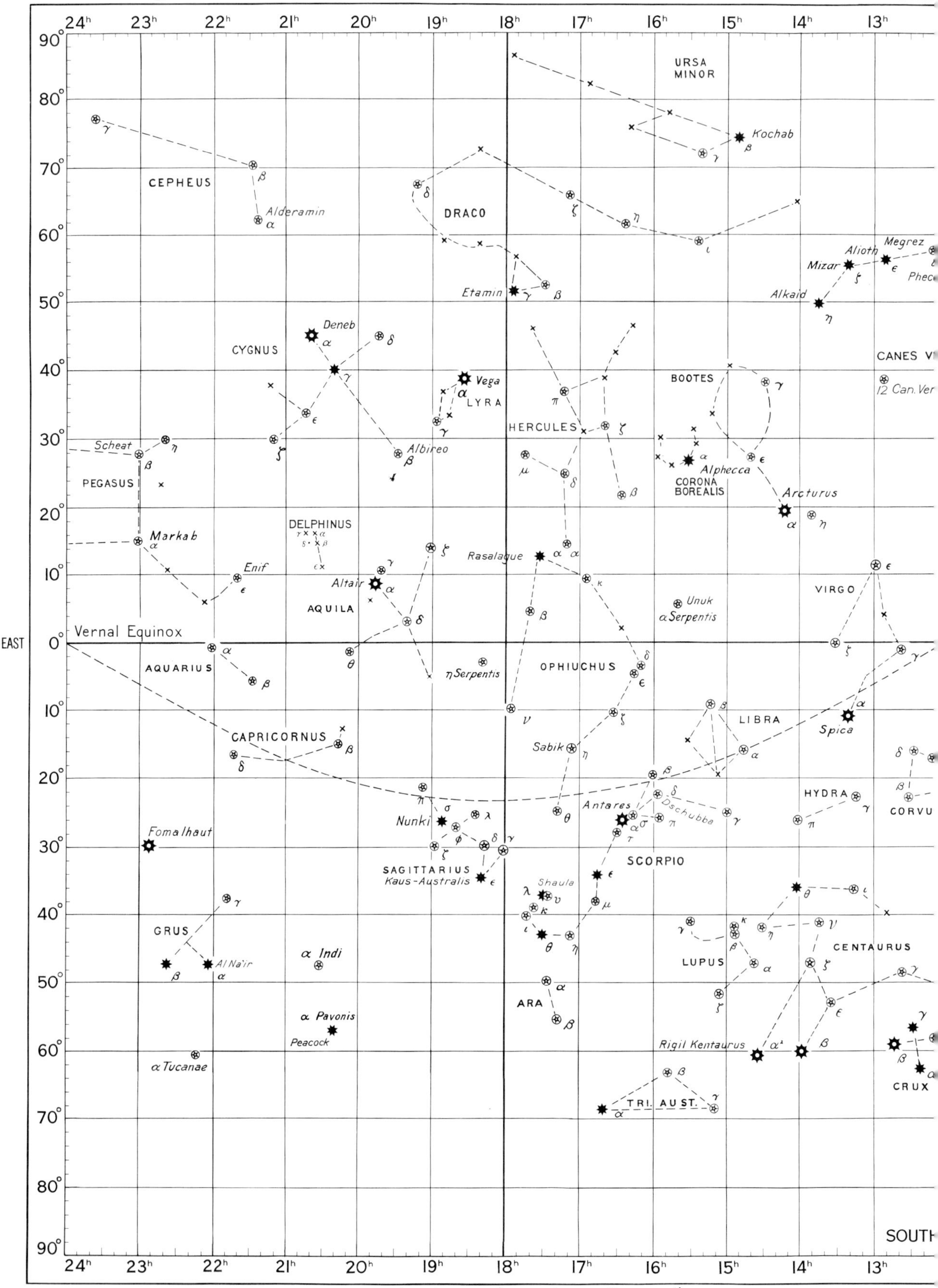

Scale of magnitudes: 1st 2nd 3rd 4th · 5th

CHART

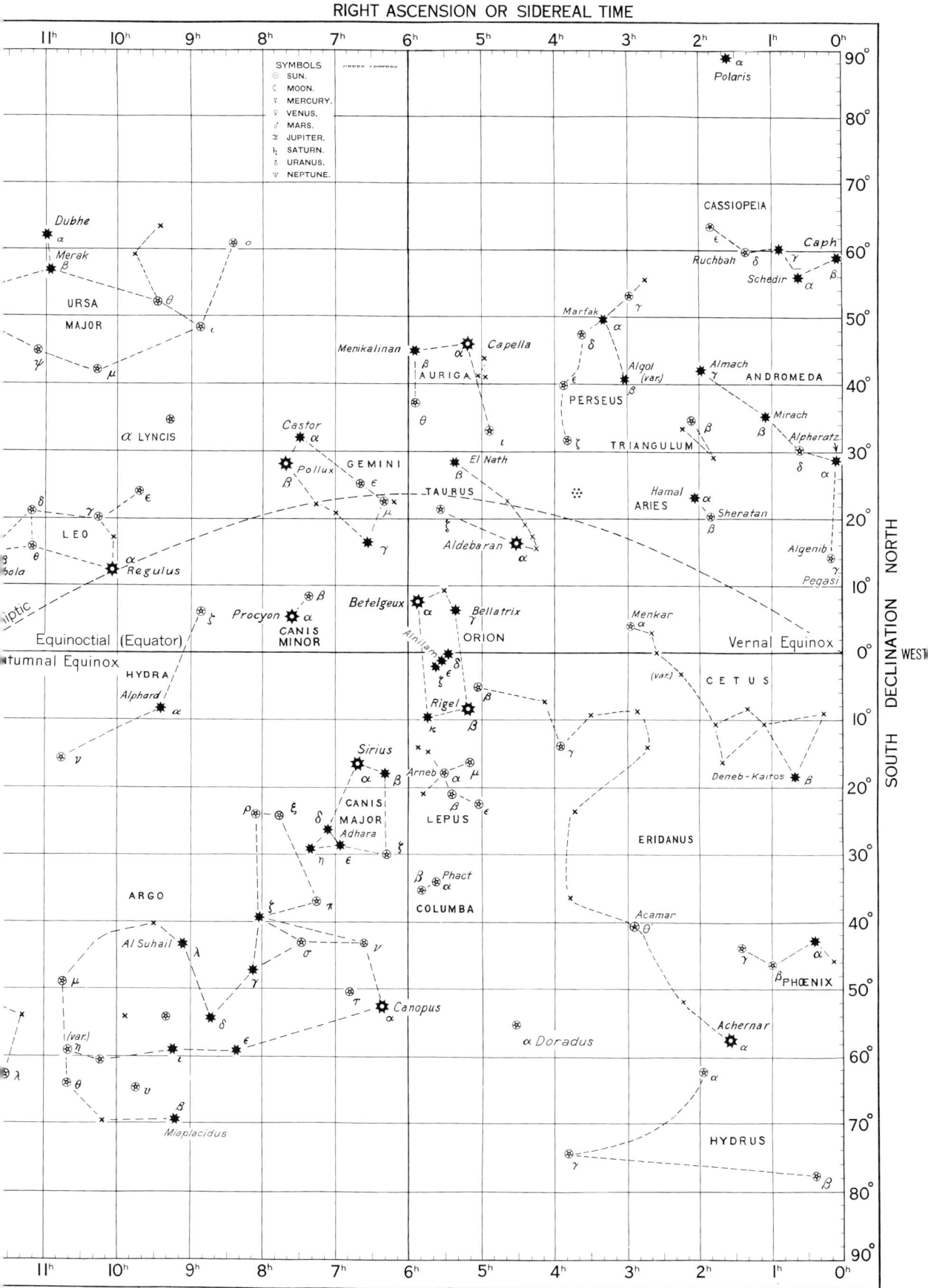

CHAPTER III

Pointing to the planets

Once you have learned how to locate and identify the constellations by pointing to them, you find the five naked-eye planets by using the same method. If you see a very bright "star" which is not on your chart, you can be sure that it is a planet. As an additional guide for identification, the planets do not twinkle, and those which are visible to the unaided eye are bright. Remember also that each planet has its distinctive color. Mars is red. Pictures taken by Viking I and sent back to Earth in July 1976 showed the surface to be the color of some of the red earth in the American western deserts. Observe that Saturn has a yellowish-red tinge and Jupiter has a silver-white color. Venus and Mercury are pearly against the twilight sky. Attention to the color will help confirm your identification. At twilight, your eyes should be able to pick out the planets before you can see the stars. In the glow of dawn, the stars disappear first and the bright light of the planets can be seen until almost sunrise.

Planets among the stars

Mercury, Venus, Mars, Jupiter and Saturn move along the path which the sun follows through the twelve constellations of the zodiac. As the planets move around the sun, we see them as if they were moving in front of the zodiacal constellations (see Figure III–1). Therefore, when giving location, we speak of a planet as being "in" one of the constellations of the zodiac. This expression has been handed down to us by past astrologers who believed that they could predict human events by combining the supposed influences of the planets with those of the constellations in which they moved at a particular time. The planets revolve around the sun—with different time schedules. Sometimes one planet appears to overtake another in its course around the zodiac. Saturn moves as if it were on a freight train schedule. Mercury, the

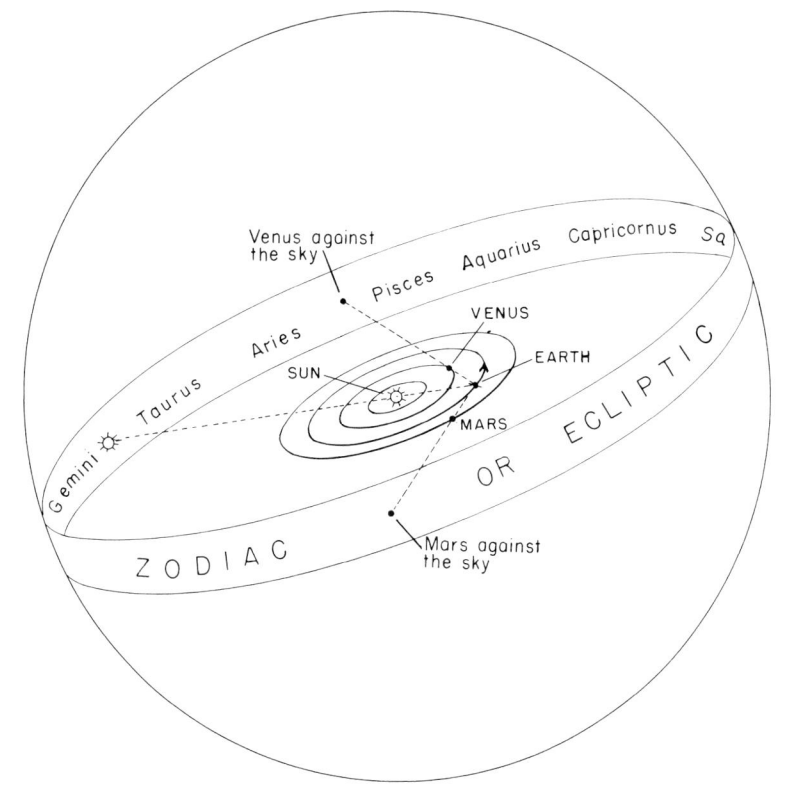

Fig. III-1. A space view showing why we see the planets along the zodiac.

fastest, lives up to its name, that is, god of speed, and moves on an express schedule.

Calendar of pointing directions

We have prepared yearly calendars (pages 84–86) giving you a month-by-month position of the planets in each of the twelve zodiacal constellations. You may use this calendar in two ways: (1) to find the constellation in which to look for a planet, and (2) to find which planet, or planets, are in a given constellation. For example, suppose the date is June 1968, and you want to point to Jupiter. On the 1968 calendar, look down the June column for Jupiter. When you find it, look over to the left-hand column for the zodiacal constellation. This happens to be Leo. Turn to page 40 in Chapter II for the pointing direction to Leo the Lion. Locating the position of Leo, you can easily point to Jupiter. You can identify it by the planet's characteristics, and by observing that there is no star at that position on maps of Leo.

In the second example, you want to know what planet is in Aquarius for January 1970. Once more refer to the calendar for 1970 and in the left-hand column find Aquarius. Then look to the right until you are under the January column. You will find that Aquarius has the planet Mars for the month of January.

The calendar has the names of the planets printed in heavy and light type. The heavy type denotes the planets which are visible or best seen in the evening hours; the light type, the morning hours. When the name of the planet is enclosed in parentheses, this object is visible only in the early evening or before dawn for part of the month.

After you have identified a planet and have followed it for some weeks or months, you may discover that it has moved along the zodiac, or ecliptic. Jupiter and Saturn are so far from the sun that they appear to move very slowly against the sky background, but Mars, Venus and Mercury move much more rapidly and you will be able to notice their changing positions among the stars. If you follow one long enough, you may discover that it appears to stand still for some nights and then start moving in the opposite direction. Let us say that early one evening you first saw Mars in Libra just to the right (west) of Antares. Then, night after night, you went out and watched this bright red planet slowly pass to the left (east) of Antares. You might expect Mars to keep on going from Libra toward Scorpius. Instead, Mars seems to pause for a night or so, and then swings toward the west. That is, it moves back to where it was a month or so before. Mars may backtrack for several months before it proceeds in its easterly direction again.

Mercury is the most difficult naked-eye planet to observe; since it is never far from the sun's direction it can be seen only in the glow of the twilight sky (Figure III-2). We have indicated on the calendar the dates around which Mercury can be seen. It will be visible only during the twilight glow for several days around the time indicated. Most people have never seen Mercury.

Venus also never appears far from the sun (Figure III-2). Therefore you can never see it during the middle of the night. When Venus is above the horizon in the twilight sky, it is the brightest object except for the moon. It is often called the "evening star," or the "morning star," although you now know that it is a planet.

The true motions of the planets

You may wonder why the planets' paths all lie along the zodiac and why they sometimes appear to change their courses and travel in the opposite direction. The planets, including our earth, revolve around the sun. All the planets and sun lie nearly in the same plane, so that they all are seen projected along almost the same background path. We may describe the motions of the planets by comparing their apparent motions, which are earth views, to their actual motions, which are space

Fig. III-2. The apparent paths of Venus and Mercury. The path of Venus is shown over a number of weeks from the time it sets just after the sun until it reaches its maximum height in the sky after dark. Thereafter, it sinks a little lower in the sky each night until it is lost in the glare of the setting sun.

The path of Mercury is similar, but Mercury can be seen only for several nights at a time at its maximum height in the sky, before and after which it is lost in the glare of the sun.

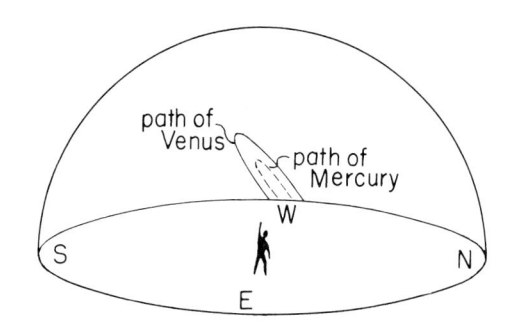

views, as shown in Figure III-1. The farther a planet is from the sun, the longer it takes to make one revolution. For example, Mercury is less than one-half the earth's distance to the sun, and it revolves around the sun in three months; whereas Saturn, the most distant naked-eye planet, is about ten times the earth-sun distance and it takes Saturn over twenty-nine years to complete its trip around the sun. Because the planets have different periods of revolution, the earth overtakes the inner planets and lags behind the outer ones in their courses around the sun. Figure III-3 shows how the switch in the direction of the path of a planet against the sky background can come about.

Observing the planets

Once you have learned how to point to the planets, you may want to study them in the same way that you studied the constellations and stars in Chapter II. The table on page 87 gives a brief selected list of interesting items to observe with the naked eye, with binoculars, or telescope.

Fig. III-3. Path of planet against background stars as seen from the earth.

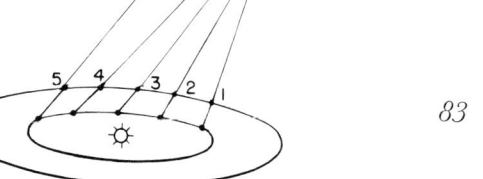

83

The Position of the Planets in the Constellation of the Zodiac
1977

	January	February	March	April	May	June	July	August	September	October	November	December
Aries	**Jupiter**	**Jupiter**				Venus Mars						
Taurus		**Jupiter**	**Jupiter**	**Jupiter**	**Jupiter**		Mars Venus Jupiter	Mars Jupiter				
Gemini			**Saturn** Saturn					Jupiter Venus	Jupiter Mars	Mars Jupiter	Jupiter	**Jupiter** Jupiter
Cancer	**Saturn** Saturn	**Saturn** Saturn		**Saturn**	**Saturn**	**Saturn**			Venus	Mars	Mars	**Mars** Mars
Leo									(Saturn) Venus	Saturn (Venus)	Saturn	Saturn
Virgo										(Venus)	(Venus)	
Libra											(Venus)	
Scorpius												
Sagittarius												
Capricornus												
Aquarius	**Venus**			(Mars)								
Pisces		**Venus**	**Venus**	(Mars)	(Mars) (Venus)							

( ) means that the sky may not be dark enough to see the constellation.

The Position of the Planets in the Constellation of the Zodiac
1978

	January	February	March	April	May	June	July	August	September	October	November	December
Aries				Venus								
Taurus	Jupiter	Jupiter	Jupiter	Venus	Venus							
Gemini	Jupiter	Mars	Mars	Jupiter	(Venus) Jupiter	(Venus) (Jupiter)						
Cancer	**Mars** Mars	Mars		Mars	Mars	(Venus)		(Jupiter)	Jupiter	Jupiter	Jupiter	Jupiter
Leo		Saturn Saturn	Saturn Saturn	Saturn Saturn	Mars Saturn	Mars Saturn	(Venus) Mars (Saturn)			(Saturn)	Saturn	Saturn
Virgo							Mars	(Venus) (Mars)	(Venus)			
Libra											(Venus)	Venus
Scorpius												
Sagittarius												
Capricornus												
Aquarius		(Venus)										
Pisces				(Venus)								

( ) means that the sky may not be dark enough to see the constellation.

The Position of the Planets in the Constellation of the Zodiac
1979

	January	February	March	April	May	June	July	August	September	October	November	December
Aries					(Mars) (Venus)	Mars						
Taurus						Mars Venus	Mars	Mars				
Gemini							(Venus)	Mars	Mars			
Cancer	Jupiter	Jupiter Jupiter	Jupiter	Jupiter	Jupiter	Jupiter	(Jupiter)		Mars	Mars		
Leo	Saturn	Saturn Saturn	Saturn	Saturn	Saturn	Saturn	Saturn		Jupiter	Mars Jupiter	Mars Jupiter	Mars Jupiter
Virgo										Saturn	Saturn	Saturn
Libra												
Scorpius	Venus										(Venus)	
Sagittarius		(Venus)										(Venus)
Capricornus												(Venus)
Aquarius												
Pisces				(Venus)	(Venus)							

( ) means that the sky may not be dark enough to see the constellation.

The Position of the Planets in the Constellation of the Zodiac
1980

	January	February	March	April	May	June	July	August	September	October	November	December
Aries			Venus									
Taurus				Venus	Venus		Venus					
Gemini					Venus	(Venus)		Venus				
Cancer									Venus			
Leo	Jupiter Mars	Jupiter Jupiter Mars Mars	Jupiter Jupiter Mars Mars	Mars Jupiter	Mars Jupiter	Mars Jupiter	Jupiter	(Jupiter)	Venus	Venus Jupiter	Jupiter	
Virgo	Saturn	Saturn Saturn	Saturn Saturn	Saturn	Saturn	Saturn Mars	Mars (Saturn)	(Mars)		Venus Jupiter (Saturn)	Venus Saturn Jupiter	Saturn Jupiter
Libra												(Venus)
Scorpius												(Venus)
Sagittarius												
Capricornus	Venus											
Aquarius	Venus	Venus										
Pisces		Venus	Venus									

( ) means that the sky may not be dark enough to see the constellation.

The Position of the Planets in the Constellation of the Zodiac
1981

	January	February	March	April	May	June	July	August	September	October	November	December
Aries												
Taurus												
Gemini							(Mars)	Mars				
Cancer									Mars			
Leo							(Venus)	(Venus)		Mars	Mars	
Virgo		Jupiter	Jupiter	Jupiter **Jupiter** Saturn **Saturn**	Jupiter **Jupiter** Saturn **Saturn**	**Jupiter Saturn**	**Jupiter Saturn**	(Venus) (Jupiter) **Saturn**	(Venus)	(Jupiter) (Saturn)	Jupiter Saturn	Mars Jupiter Saturn
Libra										(Venus)	(Venus)	Jupiter
Scorpius	Venus									Venus		
Sagittarius	Venus	(Venus)									Venus	
Capricornus	**Mars**	(Venus)										Venus
Aquarius	**Mars**	(**Mars**)	(Venus)									
Pisces												

( ) means that the sky may not be dark enough to see the constellation.

When to look for Mercury in the Twilight Sky
This planet can be seen for only a few days around the dates given below

	Evening Above the place the sun set			**Morning** Above the place the sun will rise		
1977	Apr. 19*,	Aug. 17,	Dec. 8	Jan. 22,	June 2,	Sept. 21*
1978	Apr. 3*,	July 20,	Nov. 15	May 15,	Sept. 8*,	Dec. 25*
1979	Mar. 15*,	July 12,	Nov. 1	Apr. 15,	Aug. 17*,	Dec. 5*
1980	Feb. 26*,	June 26,	Oct. 12	Mar. 26,	Aug. 1,	Nov. 14*
1981	Feb. 10*,	June 7,	Oct. 1	Mar. 13	July 15	Nov. 1*

*Those elongations which are most favorable for seeing Mercury.

The above predictions were made at the Fels Planetarium by projecting Mercury's future path on the planetarium sky.

Selected Highlights of the Naked-Eye Planets

Planet	Sketch of Telescopic View of Each Planet	With the Naked Eye	With Binoculars	With Telescopes 3-inch refractor or 6-inch reflector and larger
Mercury		Visible near sunrise or sunset for only several days at a time. Average magnitude -1.9.	No gain over the naked eye.	Phases are visible similar to those of the moon. Watch for changes in phase over a few nights. Mercury's disk is the smallest of the naked-eye planets.
Venus		Brightest object in morning or evening sky. When nearest, magnitude is -4.4; when farthest, magnitude is -3.3.	With powerful binoculars on tripod, phases may be visible. The best time to look is when it is at its brightest.	Follow the changing phases over several months. On the average, Venus appears to have 6 times the diameter of Mercury. When Venus is at its brightest, it will show a slim crescent. When faintest, it will be "full Venus."
Mars		Brightness depends on its distance from us. When nearest, magnitude is -2.7; when farthest, its magnitude is $+2.1$.	No gain over the naked eye.	Apparent size of the disk depends on the distance of Mars from the earth. When nearest to us the disk appears at its largest and then it may be possible to see its white polar caps.
Jupiter		When nearest, magnitude is -2.5. When farthest, magnitude is -1.2.	If you can hold the binoculars steady, you can see up to 4 satellites. If you count fewer, wait for a while and count again. The others may have been behind the planet in their orbits around Jupiter. They are of the sixth magnitude.	Look for the 4 bright satellites. Watch for at least an hour and you will discover that they move and revolve around Jupiter.
Saturn		Brightness varies from -0.3 to $+1.4$.	No gain over the naked eye.	Marvel at the rings. Their tilt will change over the years. Titan is the largest satellite and appears as an eighth-magnitude object. Three others may be seen with the larger telescopes.

CHAPTER IV

Pointing to the artificial satellites

Our silvery moon, topic of many romantic songs, does not need any special pointing direction in order for you to locate it. Its beauty and unique appearance cannot be ignored. We know that the moon is a member of our earth family and we say that it is our satellite in the same way that our earth and the other eight planets may be considered the satellites of the sun. These planets are known as *natural* satellites, to distinguish them from *artificial* satellites rocketed into space by man.

A few of these man-made earth satellites, destined to ride the sky for years, are sometimes visible to the naked eye; in fact, you may be able to see one as easily as you can see the brighter stars. A satellite's brightness depends on its size, how good a reflector its surface is, and its spatial relationship to the sun and the earth-bound observer. When a satellite is in the earth's shadow as illustrated in Figure IV-1 it cannot be seen. It may abruptly disappear while still high in the sky. It has passed into the earth's shadow, cut off from the rays of the sun just as a stage performer moves out of a bright spotlight. The satellite may circle the observer's sky from almost any direction, depending on the observer's position on the earth and the orbit of the satellite. These silent travelers may remain visible above the horizon for a few minutes at a time before disappearing out of view in their orbital paths. They move slowly enough so that after tentatively identifying the object you must look for a short time to see that it is moving with respect to the stars to make sure it is really an artificial satellite.

We cannot give you an easy-to-use time schedule telling you in which constellation a man-made satellite will be seen as we did for you to find the planets. Unless your local newspaper publishes the visibility of

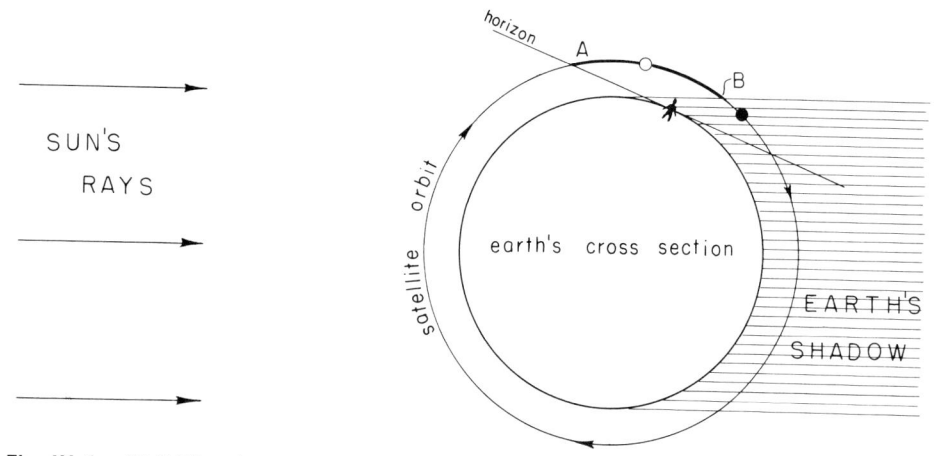

Fig. IV-1. Visibility of an artificial satellite. The little man observes the satellite rising on the western horizon at *A*. It passes over his head and disappears when it reaches *B* in its orbit because it has gone into the earth's shadow and is eclipsed.

artificial satellites along with phases of the moon in the weather section, it is chance that you will see one as it passes.

Echo I and Echo II, launched into orbits in the early 1960's, were designed to be easily visible with the naked eye. They were huge balloon-like satellites, with large reflecting surfaces. Figure IV-2 shows a photograph of Echo I racing across the sky below the triangle in Leo; it took about one minute for Echo I to move from one edge of the photograph to the other.

These objects, once set in orbit, follow the same laws as the natural satellites and planets. (See Figures IV, 3 and 4.) Some have nearly circular orbits, which means that they remain at nearly the same distance from the earth. Others move in orbits which are ellipses, which means their distances from the earth vary within one revolution. Most artificial satellites revolve around the earth in less than two hours. At best, a satellite will only be visible for a few minutes as it passes over your sky during one of its revolutions.

Fig. IV-2. Echo trail, May 9, 1962. Echo I is passing below the triangle in Leo. (Photograph with Leica, plus-X film—Joseph and Lippincott.)

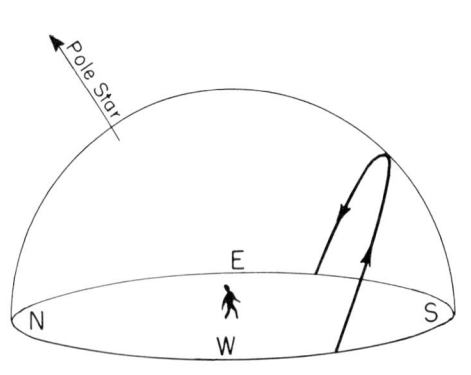

Fig. IV-3. Above, the little man sees satellite rise in southwest and set in southeast—earth view.

Fig. IV-4. Right, The satellite shown in IV-3 in its orbit—space view.

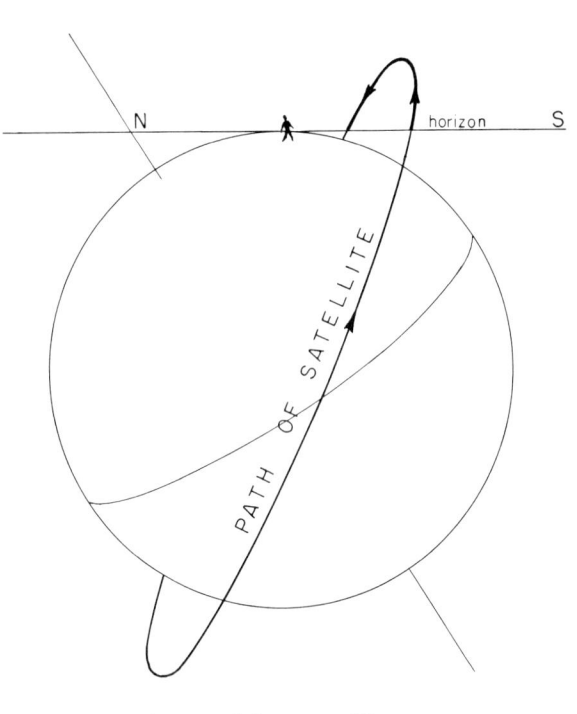

The reason why some of the man-made, earth-revolving satellites are not long-lived is that they revolve close enough to the earth to pass through the very upper parts of the atmosphere, which acts as an air brake. This causes them to slow down and dip farther into the earth's atmosphere. Finally, they spiral inward where the friction of the air action heats them so that they burn up and are destroyed. This was the fate of the Echoes after a number of years. Other satellites are designed to take another direction through the solar system, sailing on forever.

Man is studying the space surrounding our earth by sending out satellites which carry observational equipment. For example, Pegasus is a satellite designed to detect meteors—those bits of rock, iron and dust which inhabit space in our solar system. You may have seen some meteors which look like shooting stars. In general, they are only as big as grains of sand, but when they happen to drop into our atmosphere they heat up and shine like Fourth of July sparklers. Most of them burn up and change into gaseous vapors before they reach the ground. Astronomers are interested in knowing more about how much "debris" exists in the solar system. Pegasus has large protruding panels which are sensitive to the bombardment of the meteors. Because of their large reflecting surfaces, with a wing span of some 35 feet, they may appear as bright as a first or second magnitude star. Currently, a Pegasus satellite can be seen from the southern part of the United States, or farther south, closer to the equator.

In the middle and northern latitudes, the Russian Salyut and the

American Skylab may be visible from time to time. They travel in orbits which let them cross the sky from Northeast to Southwest, or Northwest to Southeast. Skylab is well-known for its variety of scientific equipment aboard. Scientists from 24 countries have directed more than 130 experiments on this spaceship. Many astronomical observations have been made and transmitted to earth for analysis. They require special antennae, transmitters and receivers to do this.

The orbiting astronomical observatory (OAO) series is primarily designed to observe with telescopes and television-like cameras the space far beyond our solar system. Information is returned to earth to a computer. Only observations which cannot be made on the earth's surface are considered. The new information supplements the vast knowledge of our universe which man continues to collect with our great earth telescopes, both optical and radio. An orbiting astronomical observatory is very difficult to see with the unaided eye, since it is thought never to be brighter than the fifth or sixth magnitude (see page 25).

One type of orbiting satellite that you cannot see, but which you are well aware of, is used for communication. The first one of its kind, the passive balloon type, was called Telstar. It was launched in July of 1962. Telstar is long since dead, but doomed to circle the earth for many decades to come. Since that time, Telstar has been supplanted by other communication satellites which allow us to share television with countries on the other side of the globe. Long distance telephone calls, across the Atlantic Ocean, for example, may also be transmitted by a satellite in orbit high above us. Satellites of this type are too small to be seen; they are generally in what is called *geo-synchronous orbit*, that is, they revolve around the earth in one day at a fixed distance of 35,000 km above the earth's surface.

Pageous is another passive balloon-type satellite currently in orbit for geodetic studies, but at this writing it is said to be in bad shape from having been hit twice by meteors.

As satellites circle the globe, we find it convenient to follow them by noting their changing positions against the background of the stars. The stars that the ancients saw and enjoyed are the same for you on earth and for the astronauts in space. People everywhere, now as in ages past, point to the stars and exclaim, "What a beautiful sight!"

GLOSSARY

Altitude The height of an object above the horizon measured in degrees of arc. This is known as an angular distance or height

Azimuth A distance around the horizon circle measured in degrees

Black hole A superdense celestial body several miles in diameter. The pull of gravity at its surface is so great that radiation cannot escape

Celestial Adjective describing something in the heavens

Circumpolar A region around the pole star in which stars never set for the observer

Constellation A group of bright stars in a small area of the sky, forming a pattern that represents a figure or object

Double star A pair of stars which are relatively close to one another and which move as a unit in space while revolving around each other

Earth view The arrangement in space of astronomical objects as seen from the earth

Eclipse When one celestial object moves in front of another cutting off our view, the more distant object is said to be in eclipse

Galaxy An enormous collection of stars. All the stars you can see with the naked eye belong to our Galaxy, which is composed of these and other stars, as well as dust and gas, concentrated in a flattened disk some 100,000 light years in diameter. Billions of other galaxies are known to exist

Horizon A great circle which is 90° from the zenith. To call the meeting place of the sky and land the horizon is therefore not always scientifically correct

Light year The distance light travels in one year at the rate of 186,000 miles per second. One light year equals about 6 trillion miles

Little man pointing system The little man *faces* a direction determined along the horizon and *points* to a position at a given altitude above the horizon

Luminous Adjective describing an object which shines by its own light. We see the stars and electric lights, etc., because they are luminous

Magnitude The degree of brightness of a celestial object or the number expressing this brightness. The larger the positive number, the fainter the object

Milky Way The luminous band in the night sky which is brightest in the direction of the constellation Sagittarius. This band indicates where the stars are densest in our Galaxy

Nebula A misty patch in the sky made of clouds of dust, gas and stars. Nebulae are concentrated along the Milky Way band. Galaxies resemble nebulae when viewed through small telescopes. This is why galaxies are often called nebulae

Orbit A path of an object in space, resulting from the gravitational pull of the object with respect to that of another nearby celestial body

Phases Changes in the appearance of the moon or other celestial objects which shine by reflected light. Many calendars give the dates for the different phases of the moon

Planet A celestial body which does not shine by its own light. We refer to the bodies which revolve around the sun as *planets*

Pole Star, or Polaris The famous second-magnitude star near the north pole of the celestial sphere

Pulsar According to present theory, a very small star of high density, thought to be the remnant of a supernova explosion. Its rapid rotation, in a fraction of a second, causes a pulsing in its brightness; therefore, these objects were named *pulsars*

Quasar Very faint celestial objects which appear different from stars or galaxies as we know them. According to one theory, quasars radiate more energy than any other objects known and are farthest from the earth

Reflection Light striking any surface which is not completely black, returns to our eyes, making it possible for us to see nonluminous objects

Satellite A natural or man-made object which revolves around the earth or any other planet

Separation The angular distance between two objects in the sky measured in degrees, minutes or seconds of arc

Standard time The time established by law or general usage for a region or country. The U.S. has four different standards of time—Eastern, Central, Mountain, and Pacific. Each of these time zones differs from the succeeding one by an hour. Some localities use "Daylight Time" in summer. This is Standard Time plus one hour

Star An astronomical object which shines by its own light or is said to be luminous. Stars are made entirely of matter in a gaseous state

Star cluster A group of stars. There are two types: galactic or open clusters and globular clusters

Star color Stars have different colors depending on their surface temperatures. The hottest stars are blue-white, while less hot stars appear orange or red. All stars, including our sun, appear orange or red near the horizon

Space view The arrangement in space of astronomical objects as seen from a position away from the earth

Telescope An instrument that collects and focuses light or radio waves for observation and analysis and is, therefore, a radiation gatherer. A "reflector" reflects light off a mirror which is shaped to focus the light where it may be photographed or viewed through an eyepiece. A "refractor" gathers and focuses the light by means of a lens

Twinkle The rapid variation of starlight caused by its passage through various layers of the earth's atmosphere

Variable stars, or Variables Stars which vary in brightness, or stars whose magnitudes change

Zenith The point in the heavens directly overhead

Zodiac The band around the celestial sphere containing 12 constellations along which the sun and the planets appear to move

INDEX

Names and page numbers of constellations are in **boldface**.

Albireo, 55
Alcor, 26, 37
Aldebaran, 29, 41
Algol, 75
Alkaid, 77
Alnitak, 24
Alpha, α, Centauri, 25
Alpheratz, 67
Al Rischa, 73
Altair, 63
Altazimuth, 22
Altitude, 20, 23, gl.
Andromeda, 66, 67
Angle, measurement of, 20, 23, 24, 26
 separation, 26, gl.
Anilam, 24
Antares, 41, 59
Aquarius, 68, 69
Aquila, 62, 63
Arcturus, 49
Argo, 78–79
Aries, 72, 73
Atmosphere, earth's, 9, 23, 89
Auriga, 42, 43
Azimuth, 20, gl.

"Beehive" (*See* Praesepe)
Bellatrix, 24
Betelgeuse, 24, 35
Big Dipper, 37
Binoculars, first use of, 27
Black hole, gl.
Boötes, 48, 49
Brightness (*See* Magnitude)

Calendar, 14
 (*See also* Planet)
Cancer, 38, 39
Canes Venatici, 78–79
Canis Major, 32, 33
Canis Minor, 32, 33
Capella, 43, 49
Capricorn, Tropic of, 69
Capricornus, 68, 69
Cassiopeia, 30, 31
Castor, 39, 77
Celestial, 8, gl.
 object, 8
 sphere, 9ff.

Centaurus, 78–79
Cepheus, 64, 65
Cetus, 78–79
Circumpolar, 11, 21, 30, 45, gl.
 constellations, 30, 36, 44, 64
Cluster, 16, 27, gl.
 double (*See* Perseus)
 globular, 57
 open, or galactic, 29
 Pleiades, Hyades, 29
 (*See also* Messier's Catalogue)
Coalsack, Northern, 55
Comet, 16, 41
Compass points, 20
Constellation, 8, 14, 15, 18, gl.
 perceiving, 18
 zodiacal, 14, 80
Coordinate system, 20–24, 76, 77
Corona Borealis, 50, 51
Corvus, 78–79
Cygnus, 54, 55

Declination, 24, 76, 77–79
Delta, δ, Cephei, 65
Demon Star (*See* Algol)
Deneb, 55
Denebola, 41
Delphinus, 62, 63
Dog Star (*See* Sirius)
Double star (*See* Star)
Draco, 78–79

Earth, 12–14
 shadow of, 91
Echo, 88, 90, 91
Eclipses, 15, 75, gl.
Ecliptic, 15, 82
El Nath, 29, 43
Equator, celestial, 10, 11, 76
 terrestrial, 11, 13, 76
Eridanus, 78–79
Eta, η, Aquilae, 63
"Evening Star," 82

Face and point, 20–23
Fomalhaut, 41, 78–79

Galaxy, 16, 17, 27, 31, gl.
  Great Nebula in Andromeda, 67
  spiral, 17, 37, 47, 67, 71, 73
  Our Milky Way, 16, 17, 61
  (*See also* Messier's Catalogue)
Ganymede, 59
Gemini, 38, 39
Gemma, 51
Great Square in Pegasus, 71
Greek alphabet, 25
Greenwich, 76
Guardians (*See* Ursa Minor)

Halley (*See* Comet)
Hamal, 73
Height (*See* Altitude)
Hercules, 56, 57
Horizon circle, 20, gl.
  (*See also* Azimuth)
Hyades (*See* Cluster)
Hydra, 78–79

Job's Coffin, 63
Jupiter, 59, 80, 82
  calendar, 84–86
  highlights, 87

Kids, The, 43

Latitude, 13, 45, 76, 77
Leo, 40, 41, 89
Lepus, 78–79
Libra, 46, 47
Light-year, 25, gl.
Little Dipper, 45
Little man pointing system, gl.
  (See also *Face and point*)
Little Milk Dipper, 61
Longitude, 76
Lyra, 52, 53, 77

Magnitude, 25, gl.
Manger (*See* Praesepe)
Markab, 71
Mars, 80, 82
  calendar, 84–86
  highlights, 87
Mercury, 80, 82, 83
  highlights, 87
  calendar, 84–86
Messier's Catalogue, 27, (M1, Crab, 29) (M2, 69) (M6, 59) (M7, 59) (M8, Lagoon, 61) (M11, 63) (M13, 57) (M15, 71) (M17, Horseshoe, 61) (M20, Trifid, 61) (M22, 61) (M24, 61) (M30, 69) (M31, Great Nebula in Andromeda, 67) (M33, 73) (M34, 75) (M35, 39) (M39, 55) (M41, 33) (M42, Orion, 35) (M44, Praesepe, 39) (M57, Ring, 53) (M67, 39) (M74, 73) (M80, 59) (M81, 37) (M92, 57) (M104, 47)
Meteor, 16
Milky Way, 16, 17, 61, gl.
  (*See also* Galaxy)
Mintaka, 24
Mizar, 26, 37
Moon, 15, 26, 53
"Morning Star," 82
Murzim, 33
Mythology, 22, 28–75

Nebula, 55, gl.
  Crab, 29
  Great Nebula in Andromeda, 67
  Horsehead, 35
  Lagoon, 61
  North American, 55
  Orion, 35
  planetary, 62
  Ring, 53
  Trifid, 61
  (*See also* Messier's Catalogue)
North Star (*See* Polaris)
Northern Cross, 55
Northern Lights, 49
Nova, 16, 29

Ophiuchus, 78–79
Orbit, 14, 88, 89, gl.
  geo-synchronous, 91
Orion, 24, 34, 35

Pegasus, 70, 71
Perseus, 74, 75
Pisces, 72, 73, 76
Planet, 15, 16, 80–83, gl.
  calendars (positions), 84–86
  highlights of, 87
  how to find, 81, 82
  mythology of, 15
  revolution around sun, 83
Pleiades (*See* Cluster)
Pointers, the, 10, 26, 37
Pointing to artificial satellites, 90
  constellations, 19–23
  planets, 80, 81
  summary of directions, 23
Polaris, Pole Star, 11, 26, 45
Pollux, 39, 77
Praesepe, 39
Procyon, 33

95

Ptolemy, 8
Pulsar, gl.

Quasar, 47, gl.

Regulus, 41
Revolution, planetary, 13, 14, 83
  (*See also* Echo)
Rigel, 24, 35
Right ascension, 24, 76

Sagittarius, 60, 61
Saiph, 24
Satellite, gl.
  artificial, 88, 89
  orbiting astronomical observatory, 91
  natural, 15, 88
  Pageous, 91
  Pegasus, 90
  Salyut, 91
  Skylab, 91
  Telstar, 90
Saturn, 63, 80, 82, 83
Schedar, 31
Scorpius, 58, 59
Serpens, 78–79
Sheratan, 73
"Shooting star" (*See* Meteor)
"Sickle" (*See* **Leo**)
Sirius, 25, 33
Sixty-one Cygni, 55
Sky dome, 8–11
Spica, 47
Star, 9, 43, gl.
  colors, 7, 8, 25, gl.
  counts, 7, 17
  clouds, 16
  distances, 9, 25
  descriptions of, 18ff
  double, 16, 26, 39, gl.
  eclipsing, 75
  magnitude or brightness, 25

names, 7, 24
nearest
  (*See* Alpha Centauri)
Star, trails, 45
  twinkle, 22
  variable, 16, 65, gl.
  white dwarf, 33
  (*See also* Sun)
Star of Bethlehem, 73
Sun, 9, 13, 14, 17, 43, 49
  corona, 51
  prominences, 49
  spots, 43, 51

Taurus, 28, 29
Telescope, 7, gl.
  first use of, 27
Time, 20–22, gl.
  schedule for planets, 84–86
Trapesium, 35
Triangulum, 78–79

Ursa Major, 36, 37
Ursa Minor, 44, 45

Variable (*See* Star)
Vega, 25, 49, 53, 77
Venus, 80, 82
  highlights, 87
Vernal equinox, 76
Viking lander, 80
Virgo, 46, 47

White dwarf (*See* Star)
World Star Chart, 76, 78–79

"Yardstick," 35

Zenith, 20, 76, 77, gl.
Zodiac, 14, 15, 80, 82, gl.
Zuben el genubi, 47
Zuben es schamali, 47